Editor: Carla Hamaguchi
Designer/Production: Moonhee Pak/Jim Lewis
Cover Designer: Barbara Peterson
Art Director: Tom Cochrane
Project Director: Carolea Williams

Table of Contents

Introduction

The *Power Practice*™ series contains dozens of ready-to-use activity pages to provide students with skill practice. The fun activities can be used to supplement and enhance what you are already teaching in your classroom. Give an activity page to students as independent class work, or send the pages home as homework to reinforce skills taught in class.

The pages in *Timed Tests: Multiplication and Division* provide students with opportunities to practice and memorize multiplication and division facts. Challenge students to do their personal best by timing them as they complete each test and keeping track of those times. The tests are intended to be repeated by students until they have reached the desired accuracy and speed.

Have students write a goal time on each test. When they complete a test, have them write their actual time on their test. Encourage students to retake a test to try to better their time and accuracy. Give students the record sheets (pages 8 and 63) to keep track of their own scores. An answer key is provided at the back of the book for quick reference.

For extra practice, have students complete the challenge pages. The multiplication challenge pages include problems with missing factors. The division pages have division facts up to 20.

Reward students for earning perfect scores, meeting their personal time goals, or mastering all the multiplication and division facts.

Give students a copy of the Multiplication Tables (see pages 4–7) as a reference for practicing the multiplication facts at home.

Use these motivating timed tests to "recharge" skill review and give students the power to succeed!

Multiplication Tables

× 2	× 3	× 4
2 × 0 = 0	3 × 0 = 0	4 × 0 = 0
2 × 1 = 2	3 × 1 = 3	4 × 1 = 4
2 × 2 = 4	3 × 2 = 6	4 × 2 = 8
2 × 3 = 6	3 × 3 = 9	4 × 3 = 12
2 × 4 = 8	3 × 4 = 12	4 × 4 = 16
2 × 5 = 10	3 × 5 = 15	4 × 5 = 20
2 × 6 = 12	3 × 6 = 18	4 × 6 = 24
2 × 7 = 14	3 × 7 = 21	4 × 7 = 28
2 × 8 = 16	3 × 8 = 24	4 × 8 = 32
2 × 9 = 18	3 × 9 = 27	4 × 9 = 36
2 × 10 = 20	3 × 10 = 30	4 × 10 = 40
2 × 11 = 22	3 × 11 = 33	4 × 11 = 44
2 × 12 = 24	3 × 12 = 36	4 × 12 = 48

Multiplication Tables

× 5	× 6	× 7
5 × 0 = 0	6 × 0 = 0	7 × 0 = 0
5 × 1 = 5	6 × 1 = 6	7 × 1 = 7
5 × 2 = 10	6 × 2 = 12	7 × 2 = 14
5 × 3 = 15	6 × 3 = 18	7 × 3 = 21
5 × 4 = 20	6 × 4 = 24	7 × 4 = 28
5 × 5 = 25	6 × 5 = 30	7 × 5 = 35
5 × 6 = 30	6 × 6 = 36	7 × 6 = 42
5 × 7 = 35	6 × 7 = 42	7 × 7 = 49
5 × 8 = 40	6 × 8 = 48	7 × 8 = 56
5 × 9 = 45	6 × 9 = 54	7 × 9 = 63
5 × 10 = 50	6 × 10 = 60	7 × 10 = 70
5 × 11 = 55	6 × 11 = 66	7 × 11 = 77
5 × 12 = 60	6 × 12 = 72	7 × 12 = 84

Multiplication Tables

× 8	× 9	× 10
8 × 0 = 0	9 × 0 = 0	10 × 0 = 0
8 × 1 = 8	9 × 1 = 9	10 × 1 = 10
8 × 2 = 16	9 × 2 = 18	10 × 2 = 20
8 × 3 = 24	9 × 3 = 27	10 × 3 = 30
8 × 4 = 32	9 × 4 = 36	10 × 4 = 40
8 × 5 = 40	9 × 5 = 45	10 × 5 = 50
8 × 6 = 48	9 × 6 = 54	10 × 6 = 60
8 × 7 = 56	9 × 7 = 63	10 × 7 = 70
8 × 8 = 64	9 × 8 = 72	10 × 8 = 80
8 × 9 = 72	9 × 9 = 81	10 × 9 = 90
8 × 10 = 80	9 × 10 = 90	10 × 10 = 100
8 × 11 = 88	9 × 11 = 99	10 × 11 = 110
8 × 12 = 96	9 × 12 = 108	10 × 12 = 120

Timed Tests: Multiplication and Division © 2004 Creative Teaching Press

Multiplication Tables

× 11	× 12
11 × 0 = 0	12 × 0 = 0
11 × 1 = 11	12 × 1 = 12
11 × 2 = 22	12 × 2 = 24
11 × 3 = 33	12 × 3 = 36
11 × 4 = 44	12 × 4 = 48
11 × 5 = 55	12 × 5 = 60
11 × 6 = 66	12 × 6 = 72
11 × 7 = 77	12 × 7 = 84
11 × 8 = 88	12 × 8 = 96
11 × 9 = 99	12 × 9 = 108
11 × 10 = 110	12 × 10 = 120
11 × 11 = 121	12 × 11 = 132
11 × 12 = 132	12 × 12 = 144

Name _____

Multiplication Record Sheet

Test	Time	Score	Test	Time	Score
×2A			Challenge ×7		
×2B			×8A		
×2C			×8B		
Challenge ×2			×8C		
×3A			Challenge ×8		
×3B			×9A		
×3C			×9B		
Challenge ×3			×9C		
Practice ×0–3A			Challenge ×9		
Practice ×0–3B			Practice ×7–9A		
×4A			Practice ×7–9B		
×4B			×10A		
×4C			×10B		
Challenge ×4			×10C		
×5A			Challenge ×10		
×5B			×11A		
×5C			×11B		
Challenge ×5			×11C		
×6A			Challenge ×11		
×6B			×12A		
×6C			×12B		
Challenge ×6			×12C		
Practice ×4–6A			Challenge ×12		
Practice ×4–6B			Practice ×10–12A		
×7A			Practice ×10–12B		
×7B			Review ×1–12A		
×7C			Review ×1–12B		

Multiplication ×2A

1 $\begin{array}{r} 2 \\ \times 8 \\ \hline \end{array}$ **2** $\begin{array}{r} 2 \\ \times 3 \\ \hline \end{array}$ **3** $\begin{array}{r} 2 \\ \times 0 \\ \hline \end{array}$ **4** $\begin{array}{r} 2 \\ \times 4 \\ \hline \end{array}$ **5** $\begin{array}{r} 2 \\ \times 6 \\ \hline \end{array}$

6 $\begin{array}{r} 2 \\ \times 12 \\ \hline \end{array}$ **7** $\begin{array}{r} 2 \\ \times 1 \\ \hline \end{array}$ **8** $\begin{array}{r} 2 \\ \times 7 \\ \hline \end{array}$ **9** $\begin{array}{r} 2 \\ \times 5 \\ \hline \end{array}$ **10** $\begin{array}{r} 2 \\ \times 2 \\ \hline \end{array}$

11 $\begin{array}{r} 2 \\ \times 5 \\ \hline \end{array}$ **12** $\begin{array}{r} 2 \\ \times 9 \\ \hline \end{array}$ **13** $\begin{array}{r} 2 \\ \times 4 \\ \hline \end{array}$ **14** $\begin{array}{r} 2 \\ \times 2 \\ \hline \end{array}$ **15** $\begin{array}{r} 2 \\ \times 11 \\ \hline \end{array}$

16 $\begin{array}{r} 2 \\ \times 1 \\ \hline \end{array}$ **17** $\begin{array}{r} 2 \\ \times 3 \\ \hline \end{array}$ **18** $\begin{array}{r} 2 \\ \times 10 \\ \hline \end{array}$ **19** $\begin{array}{r} 2 \\ \times 6 \\ \hline \end{array}$ **20** $\begin{array}{r} 2 \\ \times 7 \\ \hline \end{array}$

Timed Tests: Multiplication and Division © 2004 Creative Teaching Press

Multiplication   ×2B

1
2
× 1

2
5
× 2

3
2
× 2

4
9
× 2

5
2
× 3

6
4
× 2

7
2
× 6

8
10
× 2

9
12
× 2

10
2
× 8

11
2
×12

12
0
× 2

13
7
× 2

14
2
× 6

15
2
× 2

16
2
× 8

17
2
× 3

18
2
× 9

19
2
×11

20
2
×10

Timed Tests: Multiplication and Division © 2004 Creative Teaching Press

Goal Time _____   Actual Time _____   Score _____

Multiplication ×2C

1 2
    × 2

2 2
    × 7

3 10
    × 2

4 3
    × 2

5 2
    × 8

6 9
    × 2

7 0
    × 2

8 2
    × 5

9 1
    × 2

10 2
    × 4

11 2
    ×10

12 4
    × 2

13 2
    × 8

14 7
    × 2

15 11
    × 2

16 2
    × 5

17 2
    × 3

18 12
    × 2

19 2
    × 9

20 6
    × 2

 # Multiplication Challenge ×2

1   12
     × 2
   ‾‾‾‾

2   ☐
     × 2
   ‾‾‾‾
       0

3   2
     × 7
   ‾‾‾‾

4   ☐
     × 2
   ‾‾‾‾
      16

5   2
     ×10
   ‾‾‾‾

6   ☐
     × 2
   ‾‾‾‾
      12

7   9
     × 2
   ‾‾‾‾

8   ☐
     × 2
   ‾‾‾‾
       2

9   ☐
     × 2
   ‾‾‾‾
      22

10   ☐
     × 2
   ‾‾‾‾
      20

11   ☐
     × 2
   ‾‾‾‾
      18

12   ☐
     × 2
   ‾‾‾‾
       4

13   8
     × 2
   ‾‾‾‾

14   ☐
     × 2
   ‾‾‾‾
       8

15   ☐
     × 2
   ‾‾‾‾
      24

16   ☐
     × 2
   ‾‾‾‾
      22

17   ☐
     × 2
   ‾‾‾‾
       6

18   0
     × 2
   ‾‾‾‾

19   ☐
     × 2
   ‾‾‾‾
      14

20   12
     × 2
   ‾‾‾‾

Timed Tests: Multiplication and Division © 2004 Creative Teaching Press

Name _____ Date _____

Goal Time _____ Actual Time _____ Score _____

Multiplication ×3A

1 $\begin{array}{r} 3 \\ \times\ 2 \\ \hline \end{array}$
   2 $\begin{array}{r} 3 \\ \times\ 3 \\ \hline \end{array}$
   3 $\begin{array}{r} 3 \\ \times\ 6 \\ \hline \end{array}$
   4 $\begin{array}{r} 3 \\ \times\ 7 \\ \hline \end{array}$
   5 $\begin{array}{r} 3 \\ \times\ 9 \\ \hline \end{array}$

6 $\begin{array}{r} 3 \\ \times\ 5 \\ \hline \end{array}$
   7 $\begin{array}{r} 3 \\ \times\ 1 \\ \hline \end{array}$
   8 $\begin{array}{r} 3 \\ \times 12 \\ \hline \end{array}$
   9 $\begin{array}{r} 3 \\ \times\ 3 \\ \hline \end{array}$
   10 $\begin{array}{r} 3 \\ \times\ 0 \\ \hline \end{array}$

11 $\begin{array}{r} 3 \\ \times\ 4 \\ \hline \end{array}$
   12 $\begin{array}{r} 3 \\ \times\ 6 \\ \hline \end{array}$
   13 $\begin{array}{r} 3 \\ \times\ 7 \\ \hline \end{array}$
   14 $\begin{array}{r} 3 \\ \times 10 \\ \hline \end{array}$
   15 $\begin{array}{r} 3 \\ \times\ 2 \\ \hline \end{array}$

16 $\begin{array}{r} 3 \\ \times\ 8 \\ \hline \end{array}$
   17 $\begin{array}{r} 3 \\ \times 11 \\ \hline \end{array}$
   18 $\begin{array}{r} 3 \\ \times\ 4 \\ \hline \end{array}$
   19 $\begin{array}{r} 3 \\ \times\ 5 \\ \hline \end{array}$
   20 $\begin{array}{r} 3 \\ \times\ 1 \\ \hline \end{array}$

Name _____ Date _____

Goal Time _____ Actual Time _____ Score _____

Multiplication ×3B

1 11 × 3

2 3 × 1

3 5 × 3

4 3 × 8

5 3 × 3

6 8 × 3

7 3 × 6

8 10 × 3

9 12 × 3

10 3 × 7

11 3 × 5

12 3 × 0

13 2 × 3

14 4 × 3

15 3 × 9

16 12 × 3

17 9 × 3

18 7 × 3

19 3 × 3

20 3 × 6

Timed Tests: Multiplication and Division © 2004 Creative Teaching Press

Name _____ Date _____

Goal Time _____ Actual Time _____ Score _____

Multiplication ×3C

1 3
  × 3

2 3
  × 5

3 7
  × 3

4 3
  × 2

5 10
  × 3

6 8
  × 3

7 3
  × 4

8 11
  × 3

9 0
  × 3

10 12
  × 3

11 3
  × 7

12 1
  × 3

13 3
  × 5

14 9
  × 3

15 3
  × 3

16 6
  × 3

17 3
  × 8

18 10
  × 3

19 3
  ×12

20 11
  × 3

Multiplication Challenge ×3

1 ☐
× 3
12

2 ☐
× 3
24

3 3
× 2

4 ☐
× 3
0

5 3
× 9

6 ☐
× 3
6

7 3
× 5

8 ☐
× 3
30

9 11
× 3

10 ☐
× 3
15

11 ☐
× 3
18

12 ☐
× 3
27

13 3
× 8

14 ☐
× 3
3

15 ☐
× 3
36

16 ☐
× 3
33

17 6
× 3

18 ☐
× 3
9

19 ☐
× 3
21

20 3
× 7

Name _____ Date _____

Goal Time _____ Actual Time _____ Score _____

 # Multiplication Practice ×0–3A

1 3
     × 8

2 0
     × 9

3 3
     × 3

4 10
     × 0

5 2
     × 3

6 6
     × 0

7 8
     × 2

8 9
     × 1

9 11
     × 2

10 1
      × 3

11 5
      × 0

12 2
      × 7

13 0
      × 4

14 9
      × 3

15 7
      × 1

16 12
      × 1

17 4
      × 3

18 4
      × 1

19 2
      × 6

20 3
      × 5

Name _____ Date _____

Goal Time _____ Actual Time _____ Score _____

 # Multiplication Practice   ×0–3B

① 2
× 0

② 7
× 2

③ 0
× 8

④ 2
× 9

⑤ 2
× 3

⑥ 5
× 1

⑦ 11
× 3

⑧ 1
× 6

⑨ 2
× 2

⑩ 11
× 1

⑪ 1
× 8

⑫ 0
× 7

⑬ 2
× 8

⑭ 11
× 0

⑮ 10
× 2

⑯ 12
× 0

⑰ 3
× 7

⑱ 10
× 3

⑲ 3
× 6

⑳ 10
× 1

Timed Tests: Multiplication and Division © 2004 Creative Teaching Press

Name _____ Date _____

Goal Time _____ Actual Time _____ Score _____

Multiplication ×4A

① 4 ×5

② 4 ×9

③ 4 ×1

④ 4 ×2

⑤ 4 ×6

⑥ 4 ×12

⑦ 4 ×7

⑧ 4 ×3

⑨ 4 ×10

⑩ 4 ×0

⑪ 4 ×8

⑫ 4 ×2

⑬ 4 ×0

⑭ 4 ×7

⑮ 4 ×9

⑯ 4 ×10

⑰ 4 ×4

⑱ 4 ×12

⑲ 4 ×3

⑳ 4 ×11

Timed Tests: Multiplication and Division © 2004 Creative Teaching Press

Goal Time _____ Actual Time _____ Score _____

Multiplication ×4B

1 11
   × 4

2   4
   × 1

3   4
   × 4

4   4
   × 5

5   6
   × 4

6   4
   × 0

7   4
   × 6

8 12
   × 4

9   3
   × 4

10   7
   × 4

11   5
   × 4

12   4
   × 1

13   4
   × 7

14   8
   × 4

15   4
   × 9

16   4
   × 4

17   2
   × 4

18 10
   × 4

19 11
   × 4

20   6
   × 4

Timed Tests: Multiplication and Division © 2004 Creative Teaching Press

Name _____ Date _____

Goal Time _____ Actual Time _____ Score _____

Multiplication ×4C

❶ 0
× 4

❷ 4
×10

❸ 7
× 4

❹ 3
× 4

❺ 4
× 5

❻ 4
× 6

❼ 4
× 2

❽ 8
× 4

❾ 2
× 4

❿ 11
× 4

⓫ 4
× 9

⓬ 12
× 4

⓭ 10
× 4

⓮ 4
× 4

⓯ 5
× 4

⓰ 6
× 4

⓱ 4
× 8

⓲ 1
× 4

⓳ 4
× 0

⓴ 4
×11

Multiplication Challenge ×4

1
$\begin{array}{r} \square \\ \times\ 4 \\ \hline 16 \end{array}$

2
$\begin{array}{r} \square \\ \times\ 4 \\ \hline 24 \end{array}$

3
$\begin{array}{r} 3 \\ \times\ 4 \\ \hline \end{array}$

4
$\begin{array}{r} \square \\ \times\ 4 \\ \hline 4 \end{array}$

5
$\begin{array}{r} 4 \\ \times\ 9 \\ \hline \end{array}$

6
$\begin{array}{r} \square \\ \times\ 4 \\ \hline 44 \end{array}$

7
$\begin{array}{r} 0 \\ \times\ 4 \\ \hline \end{array}$

8
$\begin{array}{r} \square \\ \times\ 4 \\ \hline 32 \end{array}$

9
$\begin{array}{r} 10 \\ \times\ 4 \\ \hline \end{array}$

10
$\begin{array}{r} \square \\ \times\ 4 \\ \hline 8 \end{array}$

11
$\begin{array}{r} \square \\ \times\ 4 \\ \hline 40 \end{array}$

12
$\begin{array}{r} \square \\ \times\ 4 \\ \hline 48 \end{array}$

13
$\begin{array}{r} 2 \\ \times\ 4 \\ \hline \end{array}$

14
$\begin{array}{r} \square \\ \times\ 4 \\ \hline 20 \end{array}$

15
$\begin{array}{r} \square \\ \times\ 4 \\ \hline 0 \end{array}$

16
$\begin{array}{r} \square \\ \times\ 4 \\ \hline 36 \end{array}$

17
$\begin{array}{r} 4 \\ \times\ 5 \\ \hline \end{array}$

18
$\begin{array}{r} \square \\ \times\ 4 \\ \hline 12 \end{array}$

19
$\begin{array}{r} \square \\ \times\ 4 \\ \hline 28 \end{array}$

20
$\begin{array}{r} 12 \\ \times\ 4 \\ \hline \end{array}$

Timed Tests: Multiplication and Division © 2004 Creative Teaching Press

Goal Time _____ Actual Time _____ Score _____

Multiplication ×5A

1 5
  × 2

2 5
  × 8

3 5
  × 6

4 5
  × 4

5 5
  × 9

6 5
  × 7

7 5
  ×10

8 5
  × 0

9 5
  ×11

10 5
  × 3

11 5
  × 5

12 5
  × 9

13 5
  ×12

14 5
  × 3

15 5
  × 8

16 5
  × 0

17 5
  × 1

18 5
  × 7

19 5
  × 6

20 5
  ×10

Goal Time _____ Actual Time _____ Score _____

Multiplication ×5B

1 $\begin{array}{r} 5 \\ \times\ 5 \\ \hline \end{array}$
  2 $\begin{array}{r} 9 \\ \times\ 5 \\ \hline \end{array}$
  3 $\begin{array}{r} 2 \\ \times\ 5 \\ \hline \end{array}$
  4 $\begin{array}{r} 5 \\ \times\ 3 \\ \hline \end{array}$
  5 $\begin{array}{r} 12 \\ \times\ 5 \\ \hline \end{array}$

6 $\begin{array}{r} 5 \\ \times\ 1 \\ \hline \end{array}$
  7 $\begin{array}{r} 5 \\ \times 11 \\ \hline \end{array}$
  8 $\begin{array}{r} 7 \\ \times\ 5 \\ \hline \end{array}$
  9 $\begin{array}{r} 5 \\ \times\ 4 \\ \hline \end{array}$
  10 $\begin{array}{r} 10 \\ \times\ 5 \\ \hline \end{array}$

11 $\begin{array}{r} 4 \\ \times\ 5 \\ \hline \end{array}$
  12 $\begin{array}{r} 6 \\ \times\ 5 \\ \hline \end{array}$
  13 $\begin{array}{r} 5 \\ \times\ 9 \\ \hline \end{array}$
  14 $\begin{array}{r} 5 \\ \times\ 5 \\ \hline \end{array}$
  15 $\begin{array}{r} 8 \\ \times\ 5 \\ \hline \end{array}$

16 $\begin{array}{r} 11 \\ \times\ 5 \\ \hline \end{array}$
  17 $\begin{array}{r} 2 \\ \times\ 5 \\ \hline \end{array}$
  18 $\begin{array}{r} 5 \\ \times 12 \\ \hline \end{array}$
  19 $\begin{array}{r} 5 \\ \times\ 6 \\ \hline \end{array}$
  20 $\begin{array}{r} 1 \\ \times\ 5 \\ \hline \end{array}$

Timed Tests: Multiplication and Division © 2004 Creative Teaching Press

Goal Time _____ Actual Time _____ Score _____

Multiplication ×5C

1 10
×5

2 4
×5

3 12
×5

4 8
×5

5 5
×3

6 5
×7

7 5
×5

8 5
×2

9 5
×9

10 5
×6

11 11
×5

12 1
×5

13 5
×8

14 5
×12

15 6
×5

16 5
×3

17 7
×5

18 2
×5

19 5
×0

20 9
×5

Name _____   Date _____

Goal Time _____   Actual Time _____   Score _____

 # Multiplication Challenge ×5

① ☐
× 5
‾‾‾‾
45

② 5
× 8
‾‾‾‾

③ ☐
× 5
‾‾‾‾
55

④ 10
× 5
‾‾‾‾

⑤ ☐
× 5
‾‾‾‾
25

⑥ ☐
× 5
‾‾‾‾
50

⑦ ☐
× 5
‾‾‾‾
15

⑧ 11
× 5
‾‾‾‾

⑨ ☐
× 5
‾‾‾‾
60

⑩ ☐
× 5
‾‾‾‾
0

⑪ ☐
× 5
‾‾‾‾
10

⑫ 5
× 6
‾‾‾‾

⑬ ☐
× 5
‾‾‾‾
20

⑭ 5
× 9
‾‾‾‾

⑮ ☐
× 5
‾‾‾‾
35

⑯ ☐
× 5
‾‾‾‾
30

⑰ 7
× 5
‾‾‾‾

⑱ ☐
× 5
‾‾‾‾
40

⑲ ☐
× 5
‾‾‾‾
5

⑳ 5
× 2
‾‾‾‾

Multiplication   ×6A

1 $\begin{array}{r} 6 \\ \times\ 1 \\ \hline \end{array}$ **2** $\begin{array}{r} 6 \\ \times 10 \\ \hline \end{array}$ **3** $\begin{array}{r} 6 \\ \times\ 4 \\ \hline \end{array}$ **4** $\begin{array}{r} 6 \\ \times\ 8 \\ \hline \end{array}$ **5** $\begin{array}{r} 6 \\ \times\ 3 \\ \hline \end{array}$

6 $\begin{array}{r} 6 \\ \times\ 8 \\ \hline \end{array}$ **7** $\begin{array}{r} 6 \\ \times\ 5 \\ \hline \end{array}$ **8** $\begin{array}{r} 6 \\ \times\ 9 \\ \hline \end{array}$ **9** $\begin{array}{r} 6 \\ \times\ 2 \\ \hline \end{array}$ **10** $\begin{array}{r} 6 \\ \times 11 \\ \hline \end{array}$

11 $\begin{array}{r} 6 \\ \times\ 9 \\ \hline \end{array}$ **12** $\begin{array}{r} 6 \\ \times\ 1 \\ \hline \end{array}$ **13** $\begin{array}{r} 6 \\ \times 11 \\ \hline \end{array}$ **14** $\begin{array}{r} 6 \\ \times 10 \\ \hline \end{array}$ **15** $\begin{array}{r} 6 \\ \times\ 6 \\ \hline \end{array}$

16 $\begin{array}{r} 6 \\ \times\ 3 \\ \hline \end{array}$ **17** $\begin{array}{r} 6 \\ \times\ 6 \\ \hline \end{array}$ **18** $\begin{array}{r} 6 \\ \times\ 7 \\ \hline \end{array}$ **19** $\begin{array}{r} 6 \\ \times\ 4 \\ \hline \end{array}$ **20** $\begin{array}{r} 6 \\ \times 12 \\ \hline \end{array}$

Timed Tests: Multiplication and Division © 2004 Creative Teaching Press

Multiplication ×6B

1 6
× 6

2 10
× 6

3 2
× 6

4 6
×12

5 7
× 6

6 1
× 6

7 6
× 9

8 4
× 6

9 5
× 6

10 11
× 6

11 8
× 6

12 3
× 6

13 6
× 7

14 6
× 6

15 6
× 0

16 6
× 5

17 12
× 6

18 6
× 1

19 9
× 6

20 6
× 4

Timed Tests: Multiplication and Division © 2004 Creative Teaching Press

Goal Time _____   Actual Time _____   Score _____

Multiplication   ×6C

1 $\begin{array}{r} 3 \\ \times\ 6 \\ \hline \end{array}$
2 $\begin{array}{r} 11 \\ \times\ 6 \\ \hline \end{array}$
3 $\begin{array}{r} 6 \\ \times\ 5 \\ \hline \end{array}$
4 $\begin{array}{r} 6 \\ \times 12 \\ \hline \end{array}$
5 $\begin{array}{r} 9 \\ \times\ 6 \\ \hline \end{array}$

6 $\begin{array}{r} 8 \\ \times\ 6 \\ \hline \end{array}$
7 $\begin{array}{r} 6 \\ \times\ 1 \\ \hline \end{array}$
8 $\begin{array}{r} 2 \\ \times\ 6 \\ \hline \end{array}$
9 $\begin{array}{r} 7 \\ \times\ 6 \\ \hline \end{array}$
10 $\begin{array}{r} 6 \\ \times\ 0 \\ \hline \end{array}$

11 $\begin{array}{r} 6 \\ \times\ 6 \\ \hline \end{array}$
12 $\begin{array}{r} 10 \\ \times\ 6 \\ \hline \end{array}$
13 $\begin{array}{r} 4 \\ \times\ 6 \\ \hline \end{array}$
14 $\begin{array}{r} 6 \\ \times\ 3 \\ \hline \end{array}$
15 $\begin{array}{r} 6 \\ \times\ 2 \\ \hline \end{array}$

16 $\begin{array}{r} 6 \\ \times\ 4 \\ \hline \end{array}$
17 $\begin{array}{r} 12 \\ \times\ 6 \\ \hline \end{array}$
18 $\begin{array}{r} 6 \\ \times\ 8 \\ \hline \end{array}$
19 $\begin{array}{r} 5 \\ \times\ 6 \\ \hline \end{array}$
20 $\begin{array}{r} 10 \\ \times\ 6 \\ \hline \end{array}$

Name _____ Date _____

Goal Time _____ Actual Time _____ Score _____

Multiplication Challenge ×6

①
$$\begin{array}{r} \square \\ \times\ 6 \\ \hline 54 \end{array}$$

②
$$\begin{array}{r} 6 \\ \times\ 5 \\ \hline \end{array}$$

③
$$\begin{array}{r} \square \\ \times\ 6 \\ \hline 12 \end{array}$$

④
$$\begin{array}{r} 11 \\ \times\ 6 \\ \hline \end{array}$$

⑤
$$\begin{array}{r} \square \\ \times\ 6 \\ \hline 36 \end{array}$$

⑥
$$\begin{array}{r} \square \\ \times\ 6 \\ \hline 72 \end{array}$$

⑦
$$\begin{array}{r} \square \\ \times\ 6 \\ \hline 24 \end{array}$$

⑧
$$\begin{array}{r} 6 \\ \times\ 7 \\ \hline \end{array}$$

⑨
$$\begin{array}{r} \square \\ \times\ 6 \\ \hline 66 \end{array}$$

⑩
$$\begin{array}{r} \square \\ \times\ 6 \\ \hline 6 \end{array}$$

⑪
$$\begin{array}{r} \square \\ \times\ 6 \\ \hline 48 \end{array}$$

⑫
$$\begin{array}{r} 5 \\ \times\ 6 \\ \hline \end{array}$$

⑬
$$\begin{array}{r} \square \\ \times\ 6 \\ \hline 30 \end{array}$$

⑭
$$\begin{array}{r} 6 \\ \times\ 4 \\ \hline \end{array}$$

⑮
$$\begin{array}{r} \square \\ \times\ 6 \\ \hline 18 \end{array}$$

⑯
$$\begin{array}{r} \square \\ \times\ 6 \\ \hline 60 \end{array}$$

⑰
$$\begin{array}{r} 6 \\ \times\ 9 \\ \hline \end{array}$$

⑱
$$\begin{array}{r} \square \\ \times\ 6 \\ \hline 0 \end{array}$$

⑲
$$\begin{array}{r} \square \\ \times\ 6 \\ \hline 42 \end{array}$$

⑳
$$\begin{array}{r} 6 \\ \times\ 3 \\ \hline \end{array}$$

Timed Tests: Multiplication and Division © 2004 Creative Teaching Press

Name _____ Date _____

Goal Time _____ Actual Time _____ Score _____

 # Multiplication Practice   ×4–6A

1 4
×6

2 8
×5

3 5
×9

4 6
×6

5 4
×5

6 3
×4

7 7
×5

8 11
×6

9 4
×9

10 12
×4

11 5
×2

12 9
×6

13 6
×8

14 3
×6

15 8
×4

16 7
×6

17 2
×4

18 0
×6

19 7
×4

20 5
×5

Name _____   Date _____

Goal Time _____   Actual Time _____   Score _____

 **Multiplication Practice**   ×4–6B

1 5
   × 3

2 6
   × 4

3 5
   × 6

4 5
   × 9

5 10
   × 5

6 4
   × 4

7 10
   × 6

8 11
   × 4

9 6
   × 7

10 1
   × 4

11 8
   × 4

12 12
   × 6

13 6
   × 2

14 6
   × 6

15 11
   × 5

16 4
   × 5

17 6
   × 1

18 5
   × 5

19 12
   × 5

20 6
   × 9

Timed Tests: Multiplication and Division © 2006 Creative Teaching Press

Multiplication ×7A

1 7
×4

2 7
×8

3 7
×12

4 7
×2

5 7
×7

6 7
×6

7 7
×10

8 7
×3

9 7
×4

10 7
×0

11 7
×3

12 7
×2

13 7
×8

14 7
×5

15 7
×9

16 7
×7

17 7
×5

18 7
×6

19 7
×11

20 7
×10

Multiplication ×7B

① 5
  × 7

② 10
  × 7

③ 7
  × 3

④ 2
  × 7

⑤ 12
  × 7

⑥ 7
  × 7

⑦ 7
  × 0

⑧ 4
  × 7

⑨ 9
  × 7

⑩ 7
  × 8

⑪ 7
  × 3

⑫ 11
  × 7

⑬ 7
  × 5

⑭ 6
  × 7

⑮ 7
  × 1

⑯ 7
  × 9

⑰ 8
  × 7

⑱ 0
  × 7

⑲ 1
  × 7

⑳ 7
  ×11

Name _____ Date _____

Goal Time _____ Actual Time _____ Score _____

Multiplication ×7C

1 7 ×7

2 7 × 4

3 3 × 7

4 9 × 7

5 7 × 0

6 7 × 1

7 5 × 7

8 7 × 6

9 11 × 7

10 7 × 8

11 10 × 7

12 2 × 7

13 7 × 5

14 7 × 7

15 7 × 9

16 6 × 7

17 4 × 7

18 8 × 7

19 7 × 3

20 12 × 7

Multiplication Challenge ×7

1 ☐
  × 7
  ‾‾‾
  21

2 7
  × 7
  ‾‾‾

3 ☐
  × 7
  ‾‾‾
  70

4 2
  × 7
  ‾‾‾

5 ☐
  × 7
  ‾‾‾
  42

6 ☐
  × 7
  ‾‾‾
  35

7 ☐
  × 7
  ‾‾‾
  77

8 8
  × 7
  ‾‾‾

9 ☐
  × 7
  ‾‾‾
  7

10 ☐
  × 7
  ‾‾‾
  63

11 ☐
  × 7
  ‾‾‾
  56

12 9
  × 7
  ‾‾‾

13 ☐
  × 7
  ‾‾‾
  84

14 7
  × 3
  ‾‾‾

15 ☐
  × 7
  ‾‾‾
  28

16 ☐
  × 7
  ‾‾‾
  49

17 12
  × 7
  ‾‾‾

18 ☐
  × 7
  ‾‾‾
  0

19 ☐
  × 7
  ‾‾‾
  14

20 10
  × 7
  ‾‾‾

Name _____ Date _____

Goal Time _____ Actual Time _____ Score _____

Multiplication ×8A

1
$$\begin{array}{r} 8 \\ \times\ 0 \\ \hline \end{array}$$

2
$$\begin{array}{r} 8 \\ \times\ 4 \\ \hline \end{array}$$

3
$$\begin{array}{r} 8 \\ \times 12 \\ \hline \end{array}$$

4
$$\begin{array}{r} 8 \\ \times\ 9 \\ \hline \end{array}$$

5
$$\begin{array}{r} 8 \\ \times\ 2 \\ \hline \end{array}$$

6
$$\begin{array}{r} 8 \\ \times\ 8 \\ \hline \end{array}$$

7
$$\begin{array}{r} 8 \\ \times\ 1 \\ \hline \end{array}$$

8
$$\begin{array}{r} 8 \\ \times\ 3 \\ \hline \end{array}$$

9
$$\begin{array}{r} 8 \\ \times\ 6 \\ \hline \end{array}$$

10
$$\begin{array}{r} 8 \\ \times\ 5 \\ \hline \end{array}$$

11
$$\begin{array}{r} 8 \\ \times\ 3 \\ \hline \end{array}$$

12
$$\begin{array}{r} 8 \\ \times 10 \\ \hline \end{array}$$

13
$$\begin{array}{r} 8 \\ \times\ 5 \\ \hline \end{array}$$

14
$$\begin{array}{r} 8 \\ \times\ 0 \\ \hline \end{array}$$

15
$$\begin{array}{r} 8 \\ \times\ 9 \\ \hline \end{array}$$

16
$$\begin{array}{r} 8 \\ \times\ 1 \\ \hline \end{array}$$

17
$$\begin{array}{r} 8 \\ \times\ 7 \\ \hline \end{array}$$

18
$$\begin{array}{r} 8 \\ \times\ 8 \\ \hline \end{array}$$

19
$$\begin{array}{r} 8 \\ \times\ 2 \\ \hline \end{array}$$

20
$$\begin{array}{r} 8 \\ \times 11 \\ \hline \end{array}$$

Multiplication ×8B

①
$$\begin{array}{r} 8 \\ \times 10 \\ \hline \end{array}$$

②
$$\begin{array}{r} 3 \\ \times 8 \\ \hline \end{array}$$

③
$$\begin{array}{r} 6 \\ \times 8 \\ \hline \end{array}$$

④
$$\begin{array}{r} 8 \\ \times 1 \\ \hline \end{array}$$

⑤
$$\begin{array}{r} 8 \\ \times 11 \\ \hline \end{array}$$

⑥
$$\begin{array}{r} 8 \\ \times 12 \\ \hline \end{array}$$

⑦
$$\begin{array}{r} 8 \\ \times 5 \\ \hline \end{array}$$

⑧
$$\begin{array}{r} 4 \\ \times 8 \\ \hline \end{array}$$

⑨
$$\begin{array}{r} 8 \\ \times 2 \\ \hline \end{array}$$

⑩
$$\begin{array}{r} 8 \\ \times 7 \\ \hline \end{array}$$

⑪
$$\begin{array}{r} 8 \\ \times 2 \\ \hline \end{array}$$

⑫
$$\begin{array}{r} 7 \\ \times 8 \\ \hline \end{array}$$

⑬
$$\begin{array}{r} 11 \\ \times 8 \\ \hline \end{array}$$

⑭
$$\begin{array}{r} 8 \\ \times 3 \\ \hline \end{array}$$

⑮
$$\begin{array}{r} 12 \\ \times 8 \\ \hline \end{array}$$

⑯
$$\begin{array}{r} 8 \\ \times 0 \\ \hline \end{array}$$

⑰
$$\begin{array}{r} 9 \\ \times 8 \\ \hline \end{array}$$

⑱
$$\begin{array}{r} 10 \\ \times 8 \\ \hline \end{array}$$

⑲
$$\begin{array}{r} 8 \\ \times 4 \\ \hline \end{array}$$

⑳
$$\begin{array}{r} 8 \\ \times 6 \\ \hline \end{array}$$

Multiplication   ×8C

1 8
   × 3

2 4
   × 8

3 8
   × 0

4 8
   × 9

5 6
   × 8

6 5
   × 8

7 8
   × 8

8 2
   × 8

9 11
   × 8

10 8
    × 7

11 1
    × 8

12 12
    × 8

13 8
    × 10

14 3
    × 8

15 8
    × 5

16 8
    × 9

17 8
    × 6

18 8
    × 4

19 8
    × 8

20 10
    × 8

Multiplication Challenge ×8

1
```
  ☐
× 8
───
 56
```

2
```
  ☐
× 8
───
 16
```

3
```
  8
× 7
───
```

4
```
  ☐
× 8
───
 88
```

5
```
  ☐
× 8
───
 24
```

6
```
  ☐
× 8
───
 64
```

7
```
  8
× 9
───
```

8
```
  ☐
× 8
───
 32
```

9
```
  5
× 8
───
```

10
```
  ☐
× 8
───
 72
```

11
```
  ☐
× 8
───
  8
```

12
```
 12
× 8
───
```

13
```
  ☐
× 8
───
 48
```

14
```
  8
× 3
───
```

15
```
  ☐
× 8
───
 80
```

16
```
  ☐
× 8
───
 96
```

17
```
  4
× 8
───
```

18
```
  ☐
× 8
───
  0
```

19
```
  ☐
× 8
───
 40
```

20
```
  8
× 8
───
```

Goal Time _____ Actual Time _____ Score _____

Multiplication ×9A

1   9
  × 4

2   9
  × 8

3   9
  × 0

4   9
  ×11

5   9
  × 3

6   9
  ×12

7   9
  × 5

8   9
  × 2

9   9
  ×10

10   9
  × 1

11   9
  × 7

12   9
  × 4

13   9
  × 3

14   9
  × 9

15   9
  × 0

16   9
  × 8

17   9
  ×10

18   9
  × 9

19   9
  × 6

20   9
  × 5

Goal Time _____   Actual Time _____   Score _____

Multiplication   ×9B

1 9
      ×11

2 9
      × 1

3 4
      × 9

4 3
      × 9

5 9
      × 6

6 9
      × 8

7 9
      × 5

8 9
      × 2

9 9
      × 0

10 7
       × 9

11 2
       × 9

12 10
       × 9

13 12
       × 9

14 9
       × 9

15 5
       × 9

16 9
       × 7

17 9
       ×12

18 1
       × 9

19 9
       × 4

20 11
       × 9

Name _____   Date _____

Goal Time _____   Actual Time _____   Score _____

Multiplication ×9C

1 9
  × 3

2 9
  × 9

3 5
  × 9

4 10
  × 9

5 9
  × 8

6 7
  × 9

7 6
  × 9

8 9
  × 2

9 9
  × 0

10 9
  ×12

11 4
  × 9

12 11
  × 9

13 9
  × 6

14 8
  × 9

15 9
  × 7

16 9
  × 5

17 3
  × 9

18 12
  × 9

19 1
  × 9

20 9
  × 9

Multiplication Challenge ×9

1 □
 × 9
 ———
 45

2 9
 × 9
 ———

3 □
 × 9
 ———
 99

4 2
 × 9
 ———

5 □
 × 9
 ———
 36

6 □
 × 9
 ———
 72

7 10
 × 9
 ———

8 □
 × 9
 ———
 9

9 9
 × 7
 ———

10 □
 × 9
 ———
 54

11 □
 × 9
 ———
 18

12 9
 × 8
 ———

13 □
 × 9
 ———
 90

14 □
 × 9
 ———
 0

15 □
 × 9
 ———
 27

16 □
 × 9
 ———
 63

17 6
 × 9
 ———

18 □
 × 9
 ———
 108

19 □
 × 9
 ———
 81

20 9
 × 5
 ———

Timed Tests: Multiplication and Division © 2004 Creative Teaching Press

Name _____   Date _____

Goal Time _____   Actual Time _____   Score _____

 **Multiplication Practice**   ×7–9A

①   7
  × 5

②   9
  × 2

③   7
  × 4

④   8
  × 5

⑤   12
  × 7

⑥   9
  × 6

⑦   8
  × 4

⑧   10
  × 7

⑨   8
  × 6

⑩   7
  × 5

⑪   10
  × 8

⑫   11
  × 8

⑬   8
  × 9

⑭   7
  × 8

⑮   12
  × 9

⑯   7
  × 2

⑰   9
  × 4

⑱   9
  × 5

⑲   8
  × 3

⑳   10
  × 9

Name _____ Date _____

Goal Time _____ Actual Time _____ Score _____

 # Multiplication Practice ×7–9B

1 7
× 3

2 8
× 8

3 3
× 9

4 11
× 7

5 9
× 5

6 8
× 4

7 11
× 9

8 7
× 6

9 5
× 8

10 7
× 7

11 9
× 9

12 2
× 8

13 7
× 9

14 12
× 8

15 1
× 9

16 6
× 8

17 7
× 5

18 9
× 8

19 8
× 7

20 4
× 9

Timed Tests: Multiplication and Division © 2004 Creative Teaching Press

Goal Time _____ Actual Time _____ Score _____

Multiplication   ×10A

❶ $\begin{array}{r} 10 \\ \times\ 8 \\ \hline \end{array}$ **❷** $\begin{array}{r} 10 \\ \times\ 4 \\ \hline \end{array}$ **❸** $\begin{array}{r} 10 \\ \times\ 3 \\ \hline \end{array}$ **❹** $\begin{array}{r} 10 \\ \times 12 \\ \hline \end{array}$ **❺** $\begin{array}{r} 10 \\ \times\ 7 \\ \hline \end{array}$

❻ $\begin{array}{r} 10 \\ \times 10 \\ \hline \end{array}$ **❼** $\begin{array}{r} 10 \\ \times\ 1 \\ \hline \end{array}$ **❽** $\begin{array}{r} 10 \\ \times\ 0 \\ \hline \end{array}$ **❾** $\begin{array}{r} 10 \\ \times\ 2 \\ \hline \end{array}$ **❿** $\begin{array}{r} 10 \\ \times\ 5 \\ \hline \end{array}$

⓫ $\begin{array}{r} 10 \\ \times\ 6 \\ \hline \end{array}$ **⓬** $\begin{array}{r} 10 \\ \times\ 8 \\ \hline \end{array}$ **⓭** $\begin{array}{r} 10 \\ \times 11 \\ \hline \end{array}$ **⓮** $\begin{array}{r} 10 \\ \times\ 2 \\ \hline \end{array}$ **⓯** $\begin{array}{r} 10 \\ \times 10 \\ \hline \end{array}$

⓰ $\begin{array}{r} 10 \\ \times\ 5 \\ \hline \end{array}$ **⓱** $\begin{array}{r} 10 \\ \times\ 4 \\ \hline \end{array}$ **⓲** $\begin{array}{r} 10 \\ \times\ 7 \\ \hline \end{array}$ **⓳** $\begin{array}{r} 10 \\ \times\ 9 \\ \hline \end{array}$ **⓴** $\begin{array}{r} 10 \\ \times\ 3 \\ \hline \end{array}$

Goal Time _____ Actual Time _____ Score _____

Multiplication ×10B

1 10
×2

2 10
×11

3 4
×10

4 10
×9

5 10
×10

6 10
×12

7 10
×6

8 10
×8

9 10
×1

10 10
×0

11 10
×7

12 1
×10

13 2
×10

14 10
×3

15 5
×10

16 10
×4

17 10
×9

18 12
×10

19 11
×10

20 10
×6

Timed Tests: Multiplication and Division © 2004 Creative Teaching Press

Name _____ Date _____

Goal Time _____ Actual Time _____ Score _____

Multiplication ×10C

1 10
×7

2 5
×10

3 4
×10

4 10
×11

5 10
× 2

6 10
× 3

7 10
×12

8 10
× 6

9 10
× 9

10 8
×10

11 2
×10

12 10
× 9

13 12
×10

14 10
× 5

15 10
×11

16 10
×10

17 10
× 7

18 10
× 1

19 10
× 0

20 10
× 4

Goal Time _____ Actual Time _____ Score _____

 # Multiplication Challenge ×10

❶
  10
× 8
———

❷
  ☐
×10
———
 50

❸
   3
×10
———

❹
  ☐
×10
———
 90

❺
   2
×10
———

❻
  ☐
×10
———
 70

❼
  10
× 6
———

❽
  ☐
×10
———
 40

❾
  ☐
×10
———
120

❿
  ☐
×10
———
 20

⓫
  ☐
×10
———
  0

⓬
  ☐
×10
———
 10

⓭
  10
× 5
———

⓮
  ☐
×10
———
 80

⓯
  ☐
×10
———
110

⓰
  ☐
×10
———
100

⓱
  ☐
×10
———
 30

⓲
   9
×10
———

⓳
  ☐
×10
———
 60

⓴
  10
× 4
———

Timed Tests: Multiplication and Division © 2004 Creative Teaching Press

Multiplication ×11A

1 11
  × 3

2 11
  × 9

3 11
  × 7

4 11
  ×12

5 11
  × 0

6 11
  × 6

7 11
  × 5

8 11
  × 1

9 11
  ×10

10 11
   × 8

11 11
   × 2

12 11
   × 8

13 11
   ×11

14 11
   × 3

15 11
   × 4

16 11
   ×12

17 11
   × 7

18 11
   × 6

19 11
   × 9

20 11
   × 5

Goal Time _____ Actual Time _____ Score _____

Multiplication   ×11B

1 11
  × 6

2 4
  ×11

3 11
  × 1

4 11
  × 9

5 11
  × 5

6 11
  × 3

7 11
  ×11

8 8
  ×11

9 10
  ×11

10 11
  ×12

11 11
  ×10

12 5
  ×11

13 11
  × 0

14 11
  × 7

15 11
  × 2

16 11
  × 8

17 9
  ×11

18 11
  ×12

19 11
  × 4

20 11
  ×11

Timed Tests: Multiplication and Division © 2004 Creative Teaching Press

Multiplication ×11C

1 11
    × 2

2 6
    ×11

3 12
    ×11

4 11
    × 8

5 9
    ×11

6 11
    × 7

7 11
    ×10

8 1
    ×11

9 11
    × 3

10 11
     × 4

11 11
     × 5

12 11
     ×11

13 11
     ×10

14 2
     ×11

15 11
     × 7

16 3
     ×11

17 11
     × 9

18 11
     × 6

19 4
     ×11

20 11
     ×12

Goal Time _____ Actual Time _____ Score _____

 # Multiplication Challenge ×11

1 ☐
×11
66

2 ☐
×11
88

3 9
×11

4 ☐
×11
33

5 11
× 2

6 11
× 8

7 ☐
×11
55

8 4
×11

9 ☐
×11
121

10 11
× 7

11 ☐
×11
132

12 11
× 6

13 ☐
×11
22

14 ☐
×11
0

15 ☐
×11
99

16 ☐
×11
77

17 ☐
×11
11

18 11
×11

19 ☐
×11
44

20 ☐
×11
110

Timed Tests: Multiplication and Division © 2004 Creative Teaching Press

Multiplication ×12A

1 12
    × 3

2 12
    × 5

3 12
    × 0

4 12
    × 9

5 12
    × 4

6 12
    ×11

7 12
    × 2

8 12
    × 8

9 12
    × 7

10 12
     × 6

11 12
     × 9

12 12
     × 3

13 12
     × 5

14 12
     × 1

15 12
     ×10

16 12
     × 7

17 12
     × 4

18 12
     × 6

19 12
     × 8

20 12
     ×12

Name _____ Date _____

Goal Time _____ Actual Time _____ Score _____

Multiplication ×12B

1 $\begin{array}{r} 12 \\ \times\ 2 \\ \hline \end{array}$ **2** $\begin{array}{r} 12 \\ \times\ 6 \\ \hline \end{array}$ **3** $\begin{array}{r} 12 \\ \times\ 4 \\ \hline \end{array}$ **4** $\begin{array}{r} 12 \\ \times 12 \\ \hline \end{array}$ **5** $\begin{array}{r} 12 \\ \times\ 7 \\ \hline \end{array}$

6 $\begin{array}{r} 11 \\ \times 12 \\ \hline \end{array}$ **7** $\begin{array}{r} 10 \\ \times 12 \\ \hline \end{array}$ **8** $\begin{array}{r} 12 \\ \times\ 8 \\ \hline \end{array}$ **9** $\begin{array}{r} 12 \\ \times\ 7 \\ \hline \end{array}$ **10** $\begin{array}{r} 12 \\ \times\ 6 \\ \hline \end{array}$

11 $\begin{array}{r} 12 \\ \times\ 1 \\ \hline \end{array}$ **12** $\begin{array}{r} 12 \\ \times\ 5 \\ \hline \end{array}$ **13** $\begin{array}{r} 12 \\ \times\ 0 \\ \hline \end{array}$ **14** $\begin{array}{r} 12 \\ \times\ 9 \\ \hline \end{array}$ **15** $\begin{array}{r} 12 \\ \times 10 \\ \hline \end{array}$

16 $\begin{array}{r} 12 \\ \times\ 4 \\ \hline \end{array}$ **17** $\begin{array}{r} 12 \\ \times\ 2 \\ \hline \end{array}$ **18** $\begin{array}{r} 12 \\ \times\ 3 \\ \hline \end{array}$ **19** $\begin{array}{r} 12 \\ \times 12 \\ \hline \end{array}$ **20** $\begin{array}{r} 12 \\ \times 11 \\ \hline \end{array}$

Timed Tests: Multiplication and Division © 2004 Creative Teaching Press

Goal Time _____   Actual Time _____   Score _____

Multiplication ×12C

1 $\begin{array}{r} 12 \\ \times 5 \\ \hline \end{array}$ **2** $\begin{array}{r} 12 \\ \times 11 \\ \hline \end{array}$ **3** $\begin{array}{r} 3 \\ \times 12 \\ \hline \end{array}$ **4** $\begin{array}{r} 12 \\ \times 4 \\ \hline \end{array}$ **5** $\begin{array}{r} 8 \\ \times 12 \\ \hline \end{array}$

6 $\begin{array}{r} 1 \\ \times 12 \\ \hline \end{array}$ **7** $\begin{array}{r} 12 \\ \times 9 \\ \hline \end{array}$ **8** $\begin{array}{r} 12 \\ \times 10 \\ \hline \end{array}$ **9** $\begin{array}{r} 6 \\ \times 12 \\ \hline \end{array}$ **10** $\begin{array}{r} 12 \\ \times 0 \\ \hline \end{array}$

11 $\begin{array}{r} 12 \\ \times 12 \\ \hline \end{array}$ **12** $\begin{array}{r} 12 \\ \times 7 \\ \hline \end{array}$ **13** $\begin{array}{r} 11 \\ \times 12 \\ \hline \end{array}$ **14** $\begin{array}{r} 12 \\ \times 2 \\ \hline \end{array}$ **15** $\begin{array}{r} 10 \\ \times 12 \\ \hline \end{array}$

16 $\begin{array}{r} 12 \\ \times 6 \\ \hline \end{array}$ **17** $\begin{array}{r} 12 \\ \times 1 \\ \hline \end{array}$ **18** $\begin{array}{r} 12 \\ \times 4 \\ \hline \end{array}$ **19** $\begin{array}{r} 12 \\ \times 8 \\ \hline \end{array}$ **20** $\begin{array}{r} 12 \\ \times 5 \\ \hline \end{array}$

Goal Time _____   Actual Time _____   Score _____

Multiplication Challenge   ×12

1 12
    × 8

2 ☐
    ×12
    24

3   4
    ×12

4 ☐
    ×12
    60

5   3
    ×12

6 ☐
    ×12
    108

7 12
    × 9

8 ☐
    ×12
    72

9 ☐
    ×12
    144

10 ☐
    ×12
    96

11 ☐
    ×12
    132

12 12
    × 7

13 12
    × 6

14 ☐
    ×12
    36

15 ☐
    ×12
    120

16 ☐
    ×12
    48

17 ☐
    ×12
    12

18   5
    ×12

19 ☐
    ×12
    84

20 12
    ×11

Timed Tests: Multiplication and Division © 2004 Creative Teaching Press

Name _____   Date _____

Goal Time _____   Actual Time _____   Score _____

 **Multiplication Practice** ×10–12A

1 10
×6

2 8
×11

3 5
×12

4 12
× 9

5 7
×10

6 12
× 3

7 10
× 4

8 11
× 7

9 5
×10

10 12
×10

11 11
× 6

12 10
× 8

13 9
×11

14 12
× 2

15 11
× 5

16 12
× 3

17 9
×10

18 12
×12

19 11
× 3

20 6
×12

Name _____   Date _____

Goal Time _____   Actual Time _____   Score _____

 # Multiplication Practice ×10–12B

1) 12
   × 2

2) 1
   ×10

3) 11
   × 4

4) 12
   × 7

5) 9
   ×10

6) 11
   ×11

7) 12
   × 3

8) 12
   × 1

9) 10
   × 2

10) 11
   × 8

11) 3
   ×10

12) 11
   × 2

13) 12
   × 4

14) 11
   ×10

15) 12
   ×10

16) 11
   × 9

17) 10
   × 8

18) 7
   ×10

19) 12
   × 8

20) 11
   × 0

Timed Tests: Multiplication and Division © 2004 Creative Teaching Press

Name _____ Date _____

Goal Time _____ Actual Time _____ Score _____

Multiplication Review ×1–12A

1 6
      × 6

2 3
      × 4

3 8
      × 7

4 9
      × 9

5 2
      × 8

6 9
      × 3

7 11
      × 6

8 7
      × 7

9 10
      ×10

10 12
       × 5

11 11
       × 5

12 4
       × 4

13 12
       ×11

14 7
       × 2

15 8
       × 4

16 10
       × 3

17 3
       × 5

18 12
       × 6

19 6
       × 4

20 8
       ×12

Multiplication Review ×1–12B

1
$\begin{array}{r} 12 \\ \times\ 7 \\ \hline \end{array}$

2
$\begin{array}{r} 9 \\ \times\ 2 \\ \hline \end{array}$

3
$\begin{array}{r} 8 \\ \times\ 6 \\ \hline \end{array}$

4
$\begin{array}{r} 10 \\ \times\ 4 \\ \hline \end{array}$

5
$\begin{array}{r} 3 \\ \times\ 3 \\ \hline \end{array}$

6
$\begin{array}{r} 11 \\ \times\ 2 \\ \hline \end{array}$

7
$\begin{array}{r} 5 \\ \times\ 5 \\ \hline \end{array}$

8
$\begin{array}{r} 4 \\ \times\ 9 \\ \hline \end{array}$

9
$\begin{array}{r} 3 \\ \times\ 7 \\ \hline \end{array}$

10
$\begin{array}{r} 11 \\ \times\ 8 \\ \hline \end{array}$

11
$\begin{array}{r} 11 \\ \times\ 7 \\ \hline \end{array}$

12
$\begin{array}{r} 6 \\ \times 10 \\ \hline \end{array}$

13
$\begin{array}{r} 4 \\ \times\ 7 \\ \hline \end{array}$

14
$\begin{array}{r} 11 \\ \times 10 \\ \hline \end{array}$

15
$\begin{array}{r} 12 \\ \times 12 \\ \hline \end{array}$

16
$\begin{array}{r} 8 \\ \times\ 8 \\ \hline \end{array}$

17
$\begin{array}{r} 10 \\ \times\ 2 \\ \hline \end{array}$

18
$\begin{array}{r} 3 \\ \times\ 6 \\ \hline \end{array}$

19
$\begin{array}{r} 5 \\ \times\ 8 \\ \hline \end{array}$

20
$\begin{array}{r} 6 \\ \times\ 9 \\ \hline \end{array}$

Timed Tests: Multiplication and Division © 2004 Creative Teaching Press

Name _____

Division Record Sheet

Test	Time	Score	Test	Time	Score
÷2A			÷8C		
÷2B			Challenge ÷8		
÷2C			÷9A		
Challenge ÷2			÷9B		
÷3A			÷9C		
÷3B			Challenge ÷9		
÷3C			Practice ÷7–9A		
Challenge ÷3			Practice ÷7–9B		
Practice ÷1–3A			÷10A		
Practice ÷1–3B			÷10B		
÷4A			÷10C		
÷4B			Challenge ÷10		
÷4C			÷11A		
Challenge ÷4			÷11B		
÷5A			÷11C		
÷5B			Challenge ÷11		
÷5C			÷12A		
Challenge ÷5			÷12B		
÷6A			÷12C		
÷6B			Challenge ÷12		
÷6C			Practice ÷10–12A		
Challenge ÷6			Practice ÷10–12B		
Practice ÷4–6A			Review ÷1–12A		
Practice ÷4–6B			Review ÷1–12B		
÷7A			Review x and ÷ A		
÷7B			Review x and ÷ B		
÷7C			Review x and ÷ C		
Challenge ÷7			Review x and ÷ D		
÷8A			Review x and ÷ E		
÷8B			Review x and ÷ F		

Name _____ Date _____

Goal Time _____ Actual Time _____ Score _____

Division ÷2A

1 2)‾1‾2‾   **2** 2)‾4‾   **3** 2)‾1‾2‾   **4** 2)‾2‾   **5** 2)‾6‾

6 2)‾2‾0‾   **7** 2)‾2‾4‾   **8** 2)‾1‾6‾   **9** 2)‾8‾   **10** 2)‾1‾4‾

11 2)‾1‾0‾   **12** 2)‾6‾   **13** 2)‾1‾8‾   **14** 2)‾1‾2‾   **15** 2)‾2‾4‾

16 2)‾8‾   **17** 2)‾2‾2‾   **18** 2)‾4‾   **19** 2)‾1‾6‾   **20** 2)‾1‾0‾

Timed Test: Multiplication and Division © 2006 Creative Teaching Press

Division ÷2B

1 2)22   **2** 2)4   **3** 2)16   **4** 2)10   **5** 2)12

6 2)20   **7** 2)18   **8** 2)8   **9** 2)24   **10** 2)14

11 2)10   **12** 2)12   **13** 2)22   **14** 2)6   **15** 2)20

16 2)2   **17** 2)14   **18** 2)4   **19** 2)16   **20** 2)18

Goal Time _____ Actual Time _____ Score _____

Division   ÷2C

1 $2\overline{)22}$ **2** $2\overline{)4}$ **3** $2\overline{)16}$ **4** $2\overline{)10}$ **5** $2\overline{)12}$

6 $2\overline{)20}$ **7** $2\overline{)18}$ **8** $2\overline{)8}$ **9** $2\overline{)24}$ **10** $2\overline{)14}$

11 $2\overline{)2}$ **12** $2\overline{)6}$ **13** $2\overline{)10}$ **14** $2\overline{)22}$ **15** $2\overline{)4}$

16 $2\overline{)18}$ **17** $2\overline{)24}$ **18** $2\overline{)20}$ **19** $2\overline{)14}$ **20** $2\overline{)8}$

Timed Tests: Multiplication and Division © 2004 Creative Teaching Press

Division Challenge ÷2

1 2)24 **2** 2)6 **3** 2)16 **4** 2)30 **5** 2)38

6 2)20 **7** 2)34 **8** 2)28 **9** 2)4 **10** 2)14

11 2)18 **12** 2)12 **13** 2)40 **14** 2)22 **15** 2)2

16 2)8 **17** 2)36 **18** 2)10 **19** 2)32 **20** 2)26

Goal Time _____ Actual Time _____ Score _____

Division ÷3A

① 3)6   **②** 3)30   **③** 3)9   **④** 3)12   **⑤** 3)15

⑥ 3)21   **⑦** 3)27   **⑧** 3)3   **⑨** 3)18   **⑩** 3)24

⑪ 3)15   **⑫** 3)33   **⑬** 3)30   **⑭** 3)9   **⑮** 3)3

⑯ 3)36   **⑰** 3)12   **⑱** 3)18   **⑲** 3)6   **⑳** 3)27

Goal Time _____ Actual Time _____ Score _____

Division ÷3B

1 3)12   **2** 3)24   **3** 3)30   **4** 3)18   **5** 3)3

6 3)9   **7** 3)21   **8** 3)33   **9** 3)6   **10** 3)27

11 3)18   **12** 3)15   **13** 3)21   **14** 3)24   **15** 3)12

16 3)3   **17** 3)27   **18** 3)6   **19** 3)33   **20** 3)9

Division   ÷3C

1 3)3   **2** 3)15   **3** 3)33   **4** 3)21   **5** 3)27

6 3)24   **7** 3)9   **8** 3)30   **9** 3)18   **10** 3)6

11 3)12   **12** 3)6   **13** 3)27   **14** 3)30   **15** 3)15

16 3)24   **17** 3)18   **18** 3)21   **19** 3)3   **20** 3)9

Name _____   Date _____

Goal Time _____   Actual Time _____   Score _____

 # Division Challenge ÷3

1 3)24   **2** 3)54   **3** 3)48   **4** 3)33   **5** 3)18

6 3)45   **7** 3)30   **8** 3)9   **9** 3)21   **10** 3)42

11 3)15   **12** 3)60   **13** 3)51   **14** 3)36   **15** 3)3

16 3)39   **17** 3)57   **18** 3)12   **19** 3)27   **20** 3)6

Division Practice   ÷1–3A

1 2⟌8   **2** 3⟌21   **3** 1⟌4   **4** 2⟌16   **5** 1⟌8

6 1⟌5   **7** 3⟌18   **8** 2⟌20   **9** 3⟌12   **10** 1⟌7

11 1⟌12   **12** 2⟌18   **13** 3⟌15   **14** 1⟌9   **15** 3⟌21

16 2⟌24   **17** 1⟌10   **18** 3⟌9   **19** 2⟌22   **20** 2⟌14

Division Practice   ÷1–3B

1 $1\overline{)3}$ **2** $3\overline{)18}$ **3** $2\overline{)4}$ **4** $1\overline{)11}$ **5** $2\overline{)6}$

6 $3\overline{)33}$ **7** $2\overline{)10}$ **8** $1\overline{)2}$ **9** $3\overline{)36}$ **10** $1\overline{)5}$

11 $1\overline{)6}$ **12** $3\overline{)24}$ **13** $2\overline{)18}$ **14** $2\overline{)2}$ **15** $3\overline{)27}$

16 $3\overline{)6}$ **17** $2\overline{)12}$ **18** $3\overline{)30}$ **19** $1\overline{)1}$ **20** $2\overline{)22}$

Division   ÷4A

1) 4⟌16   **2)** 4⟌40   **3)** 4⟌8   **4)** 4⟌24   **5)** 4⟌32

6) 4⟌4   **7)** 4⟌28   **8)** 4⟌44   **9)** 4⟌20   **10)** 4⟌36

11) 4⟌32   **12)** 4⟌8   **13)** 4⟌12   **14)** 4⟌48   **15)** 4⟌28

16) 4⟌24   **17)** 4⟌20   **18)** 4⟌36   **19)** 4⟌16   **20)** 4⟌4

Timed Tests: Multiplication and Division © 2004 Creative Teaching Press

Division ÷4B

1 4)‾4   **2** 4)‾12   **3** 4)‾28   **4** 4)‾20   **5** 4)‾48

6 4)‾40   **7** 4)‾24   **8** 4)‾16   **9** 4)‾32   **10** 4)‾8

11 4)‾28   **12** 4)‾36   **13** 4)‾44   **14** 4)‾4   **15** 4)‾12

16 4)‾32   **17** 4)‾16   **18** 4)‾28   **19** 4)‾40   **20** 4)‾20

Goal Time _____ Actual Time _____ Score _____

Division ÷4C

1 4)20 **2** 4)48 **3** 4)24 **4** 4)12 **5** 4)36

6 4)8 **7** 4)32 **8** 4)16 **9** 4)40 **10** 4)4

11 4)40 **12** 4)28 **13** 4)12 **14** 4)44 **15** 4)24

16 4)48 **17** 4)36 **18** 4)16 **19** 4)20 **20** 4)8

Timed Tests: Multiplication and Division © 2004 Creative Teaching Press

Name _____ Date _____

Goal Time _____ Actual Time _____ Score _____

Division Challenge ÷4

1 4)64   **2** 4)24   **3** 4)72   **4** 4)40   **5** 4)56

6 4)20   **7** 4)76   **8** 4)4   **9** 4)36   **10** 4)12

11 4)60   **12** 4)44   **13** 4)80   **14** 4)28   **15** 4)68

16 4)32   **17** 4)52   **18** 4)8   **19** 4)16   **20** 4)48

Name _____ Date _____

Goal Time _____ Actual Time _____ Score _____

Division   ÷5A

1 5)‾50   **2** 5)‾10   **3** 5)‾35   **4** 5)‾20   **5** 5)‾45

6 5)‾25   **7** 5)‾5   **8** 5)‾55   **9** 5)‾30   **10** 5)‾60

11 5)‾40   **12** 5)‾45   **13** 5)‾15   **14** 5)‾10   **15** 5)‾35

16 5)‾20   **17** 5)‾30   **18** 5)‾50   **19** 5)‾25   **20** 5)‾5

Timed Tests: Multiplication and Division © 2004 Creative Teaching Press

Goal Time _____   Actual Time _____   Score _____

Division   ÷5B

① 5)5   **②** 5)40   **③** 5)30   **④** 5)15   **⑤** 5)45

⑥ 5)25   **⑦** 5)60   **⑧** 5)10   **⑨** 5)50   **⑩** 5)35

⑪ 5)20   **⑫** 5)55   **⑬** 5)25   **⑭** 5)45   **⑮** 5)5

⑯ 5)10   **⑰** 5)40   **⑱** 5)15   **⑲** 5)30   **⑳** 5)60

Division ÷5C

1 5)40   **2** 5)55   **3** 5)30   **4** 5)15   **5** 5)50

6 5)25   **7** 5)10   **8** 5)45   **9** 5)60   **10** 5)20

11 5)50   **12** 5)35   **13** 5)5   **14** 5)10   **15** 5)30

16 5)15   **17** 5)45   **18** 5)60   **19** 5)40   **20** 5)25

Name _____ Date _____

Goal Time _____ Actual Time _____ Score _____

 # Division Challenge ÷5

1 5)45   **2** 5)80   **3** 5)25   **4** 5)10   **5** 5)55

6 5)70   **7** 5)5   **8** 5)60   **9** 5)35   **10** 5)90

11 5)20   **12** 5)40   **13** 5)95   **14** 5)30   **15** 5)85

16 5)75   **17** 5)50   **18** 5)100   **19** 5)15   **20** 5)65

Name _____ Date _____

Goal Time _____ Actual Time _____ Score _____

Division ÷6A

1 6)24   **2** 6)60   **3** 6)36   **4** 6)12   **5** 6)48

6 6)6   **7** 6)54   **8** 6)18   **9** 6)66   **10** 6)30

11 6)42   **12** 6)12   **13** 6)72   **14** 6)60   **15** 6)36

16 6)18   **17** 6)48   **18** 4)54   **19** 6)24   **20** 6)6

Timed Tests: Multiplication and Division © 2004 Creative Teaching Press

Division   ÷6B

1 $6\overline{)6}$ **2** $6\overline{)54}$ **3** $6\overline{)72}$ **4** $6\overline{)18}$ **5** $6\overline{)60}$

6 $6\overline{)30}$ **7** $6\overline{)66}$ **8** $6\overline{)48}$ **9** $6\overline{)42}$ **10** $6\overline{)24}$

11 $6\overline{)36}$ **12** $6\overline{)12}$ **13** $6\overline{)60}$ **14** $6\overline{)54}$ **15** $6\overline{)72}$

16 $6\overline{)66}$ **17** $6\overline{)42}$ **18** $6\overline{)24}$ **19** $6\overline{)48}$ **20** $6\overline{)30}$

Division ÷6C

1 $6\overline{)30}$ **2** $6\overline{)18}$ **3** $6\overline{)54}$ **4** $6\overline{)42}$ **5** $6\overline{)72}$

6 $6\overline{)36}$ **7** $6\overline{)60}$ **8** $6\overline{)24}$ **9** $6\overline{)48}$ **10** $6\overline{)12}$

11 $6\overline{)6}$ **12** $6\overline{)72}$ **13** $6\overline{)66}$ **14** $6\overline{)30}$ **15** $6\overline{)18}$

16 $6\overline{)12}$ **17** $6\overline{)48}$ **18** $6\overline{)54}$ **19** $6\overline{)36}$ **20** $6\overline{)24}$

Timed Tests: Multiplication and Division © 2004 Creative Teaching Press

Division Challenge ÷6

1 6)24 **2** 6)108 **3** 6)54 **4** 6)36 **5** 6)72

6 6)66 **7** 6)12 **8** 6)90 **9** 6)114 **10** 6)30

11 6)42 **12** 6)102 **13** 6)6 **14** 6)78 **15** 6)60

16 6)84 **17** 6)48 **18** 6)120 **19** 6)18 **20** 6)96

Division Practice   ÷4–6A

1 4)16   **2** 6)6   **3** 4)44   **4** 5)15   **5** 6)66

6 6)6   **7** 5)35   **8** 6)18   **9** 4)8   **10** 5)20

11 6)30   **12** 5)55   **13** 4)12   **14** 6)54   **15** 5)5

16 5)45   **17** 4)28   **18** 5)25   **19** 4)24   **20** 6)42

Timed Tests: Multiplication and Division © 2004 Creative Teaching Press

Name _____ Date _____

Goal Time _____ Actual Time _____ Score _____

Division Practice ÷4–6B

1 $6\overline{)30}$ **2** $4\overline{)20}$ **3** $5\overline{)25}$ **4** $6\overline{)72}$ **5** $5\overline{)50}$

6 $5\overline{)60}$ **7** $6\overline{)12}$ **8** $4\overline{)36}$ **9** $5\overline{)10}$ **10** $6\overline{)36}$

11 $4\overline{)40}$ **12** $5\overline{)20}$ **13** $6\overline{)24}$ **14** $4\overline{)4}$ **15** $5\overline{)40}$

16 $6\overline{)48}$ **17** $4\overline{)20}$ **18** $5\overline{)30}$ **19** $6\overline{)60}$ **20** $4\overline{)48}$

Division ÷7A

1 7)28 **2** 7)56 **3** 7)14 **4** 7)77 **5** 7)42

6 7)35 **7** 7)7 **8** 7)49 **9** 7)21 **10** 7)63

11 7)70 **12** 7)21 **13** 7)84 **14** 7)56 **15** 7)28

16 7)77 **17** 7)42 **18** 7)35 **19** 7)7 **20** 7)49

Division   ÷7B

1 7)21 **2** 7)35 **3** 7)7 **4** 7)63 **5** 7)49

6 7)14 **7** 7)70 **8** 7)42 **9** 7)84 **10** 7)28

11 7)56 **12** 7)49 **13** 7)77 **14** 7)35 **15** 7)14

16 7)84 **17** 7)28 **18** 7)21 **19** 7)70 **20** 7)42

Division ÷7C

1 7)84 **2** 7)56 **3** 7)28 **4** 7)42 **5** 7)70

6 7)21 **7** 7)49 **8** 7)63 **9** 7)14 **10** 7)77

11 7)70 **12** 7)42 **13** 7)7 **14** 7)35 **15** 7)84

16 7)49 **17** 7)77 **18** 7)21 **19** 7)63 **20** 7)56

Goal Time _____ Actual Time _____ Score _____

Division Challenge ÷7

1 $7\overline{)7}$ **2** $7\overline{)49}$ **3** $7\overline{)119}$ **4** $7\overline{)84}$ **5** $7\overline{)28}$

6 $7\overline{)105}$ **7** $7\overline{)63}$ **8** $7\overline{)14}$ **9** $7\overline{)140}$ **10** $7\overline{)91}$

11 $7\overline{)21}$ **12** $7\overline{)70}$ **13** $7\overline{)126}$ **14** $7\overline{)35}$ **15** $7\overline{)112}$

16 $7\overline{)77}$ **17** $7\overline{)42}$ **18** $7\overline{)98}$ **19** $7\overline{)133}$ **20** $7\overline{)56}$

Name _____   Date _____

Goal Time _____   Actual Time _____   Score _____

Division ÷8A

① 8)‾4̄0̄   **②** 8)‾6̄4̄   **③** 8)‾2̄4̄   **④** 8)‾5̄6̄   **⑤** 8)‾8̄

⑥ 8)‾1̄6̄   **⑦** 8)‾8̄0̄   **⑧** 8)‾4̄8̄   **⑨** 8)‾7̄2̄   **⑩** 8)‾3̄2̄

⑪ 8)‾5̄6̄   **⑫** 8)‾2̄4̄   **⑬** 8)‾8̄8̄   **⑭** 8)‾9̄6̄   **⑮** 8)‾6̄4̄

⑯ 8)‾3̄2̄   **⑰** 8)‾4̄0̄   **⑱** 8)‾7̄2̄   **⑲** 8)‾1̄6̄   **⑳** 8)‾4̄8̄

Timed Tests: Multiplication and Division © 2004 Creative Teaching Press

Goal Time _____ Actual Time _____ Score _____

Division ÷8B

1 8)32   **2** 8)72   **3** 8)16   **4** 8)88   **5** 8)48

6 8)24   **7** 8)56   **8** 8)96   **9** 8)40   **10** 8)64

11 8)88   **12** 8)8   **13** 8)48   **14** 8)80   **15** 8)72

16 8)64   **17** 8)96   **18** 8)40   **19** 8)56   **20** 8)24

Division ÷8C

1 8$\overline{)64}$ **2** 8$\overline{)24}$ **3** 8$\overline{)80}$ **4** 8$\overline{)40}$ **5** 8$\overline{)96}$

6 8$\overline{)48}$ **7** 8$\overline{)72}$ **8** 8$\overline{)32}$ **9** 8$\overline{)88}$ **10** 8$\overline{)8}$

11 8$\overline{)56}$ **12** 8$\overline{)96}$ **13** 8$\overline{)16}$ **14** 8$\overline{)80}$ **15** 8$\overline{)64}$

16 8$\overline{)88}$ **17** 8$\overline{)40}$ **18** 8$\overline{)48}$ **19** 8$\overline{)72}$ **20** 8$\overline{)32}$

Name _____ Date _____

Goal Time _____ Actual Time _____ Score _____

Division Challenge ÷8

1 8)‾40   **2** 8)‾96   **3** 8)‾8   **4** 8)‾120   **5** 8)‾64

6 8)‾112   **7** 8)‾48   **8** 8)‾144   **9** 8)‾24   **10** 8)‾80

11 8)‾16   **12** 8)‾88   **13** 8)‾56   **14** 8)‾104   **15** 8)‾152

16 8)‾136   **17** 8)‾32   **18** 8)‾160   **19** 8)‾72   **20** 8)‾128

Division   ÷9A

1 9)63   **2** 9)9   **3** 9)45   **4** 9)81   **5** 9)27

6 9)36   **7** 9)72   **8** 9)90   **9** 9)18   **10** 9)54

11 9)27   **12** 9)45   **13** 9)108   **14** 9)63   **15** 9)99

16 9)54   **17** 9)18   **18** 9)9   **19** 9)72   **20** 9)90

Division   ÷9B

1 9)81   **2** 9)18   **3** 9)63   **4** 9)99   **5** 9)54

6 9)27   **7** 9)108   **8** 9)45   **9** 9)9   **10** 9)72

11 9)99   **12** 9)54   **13** 9)36   **14** 9)81   **15** 9)90

16 9)27   **17** 9)36   **18** 9)72   **19** 9)63   **20** 9)108

Division   ÷9C

1 9)27   **2** 9)54   **3** 9)72   **4** 9)36   **5** 9)99

6 9)45   **7** 9)81   **8** 9)63   **9** 9)18   **10** 9)108

11 9)36   **12** 9)9   **13** 9)27   **14** 9)90   **15** 9)54

16 9)108   **17** 9)81   **18** 9)99   **19** 9)45   **20** 9)63

Name _____ Date _____

Goal Time _____ Actual Time _____ Score _____

Division Challenge ÷9

1 9)27 **2** 9)108 **3** 9)153 **4** 9)9 **5** 9)72

6 9)18 **7** 9)126 **8** 9)45 **9** 9)162 **10** 9)99

11 9)90 **12** 9)36 **13** 9)171 **14** 9)63 **15** 9)117

16 9)54 **17** 9)144 **18** 9)81 **19** 9)180 **20** 9)135

Name _____ Date _____

Goal Time _____ Actual Time _____ Score _____

 # Division Practice ÷7–9A

1 7⟌42   **2** 8⟌56   **3** 9⟌45   **4** 7⟌70   **5** 9⟌81

6 9⟌54   **7** 7⟌84   **8** 8⟌40   **9** 7⟌14   **10** 8⟌72

11 8⟌24   **12** 9⟌99   **13** 7⟌56   **14** 8⟌88   **15** 9⟌27

16 9⟌63   **17** 7⟌28   **18** 8⟌32   **19** 7⟌7   **20** 8⟌64

Timed Tests: Multiplication and Division © 2004 Creative Teaching Press

Division Practice   ÷7–9B

1 8)16   **2** 9)90   **3** 8)64   **4** 7)49   **5** 9)18

6 7)63   **7** 8)80   **8** 9)108   **9** 8)8   **10** 7)21

11 7)28   **12** 9)54   **13** 8)32   **14** 7)77   **15** 9)36

16 7)35   **17** 8)96   **18** 9)72   **19** 8)48   **20** 9)72

Division   ÷10A

1 10)40   **2** 10)70   **3** 10)20   **4** 10)100   **5** 10)60

6 10)120   **7** 10)10   **8** 10)80   **9** 10)30   **10** 10)90

11 10)110   **12** 10)60   **13** 10)50   **14** 10)70   **15** 10)20

16 10)30   **17** 10)80   **18** 10)90   **19** 10)40   **20** 10)120

Division   ÷10B

1 10)30   **2** 10)60   **3** 10)110   **4** 10)90   **5** 10)120

6 10)80   **7** 10)100   **8** 10)50   **9** 10)20   **10** 10)70

11 10)40   **12** 10)70   **13** 10)10   **14** 10)30   **15** 10)110

16 10)90   **17** 10)120   **18** 10)80   **19** 10)50   **20** 10)100

Goal Time _____   Actual Time _____   Score _____

Division   ÷10C

① 10)50   **②** 10)100   **③** 10)30   **④** 10)70   **⑤** 10)80

⑥ 10)90   **⑦** 10)110   **⑧** 10)60   **⑨** 10)40   **⑩** 10)10

⑪ 10)20   **⑫** 10)80   **⑬** 10)120   **⑭** 10)30   **⑮** 10)50

⑯ 10)110   **⑰** 10)40   **⑱** 10)70   **⑲** 10)60   **⑳** 10)90

Name _____ Date _____

Goal Time _____ Actual Time _____ Score _____

 **Division Challenge ÷10**

1 10)70 **2** 10)130 **3** 10)100 **4** 10)20 **5** 10)170

6 10)120 **7** 10)10 **8** 10)80 **9** 10)150 **10** 10)40

11 10)30 **12** 10)140 **13** 10)60 **14** 10)190 **15** 10)160

16 10)110 **17** 10)180 **18** 10)50 **19** 10)200 **20** 10)90

Division   ÷11A

1 $11\overline{)55}$ **2** $11\overline{)99}$ **3** $11\overline{)11}$ **4** $11\overline{)121}$ **5** $11\overline{)77}$

6 $11\overline{)22}$ **7** $11\overline{)88}$ **8** $11\overline{)66}$ **9** $11\overline{)110}$ **10** $11\overline{)33}$

11 $11\overline{)132}$ **12** $11\overline{)77}$ **13** $11\overline{)44}$ **14** $11\overline{)121}$ **15** $11\overline{)99}$

16 $11\overline{)33}$ **17** $11\overline{)66}$ **18** $11\overline{)22}$ **19** $11\overline{)88}$ **20** $11\overline{)55}$

Timed Tests: Multiplication and Division © 2006 Creative Teaching Press

Division ÷11B

1 11)66 **2** 11)110 **3** 11)22 **4** 11)88 **5** 11)44

6 11)33 **7** 11)99 **8** 11)77 **9** 11)132 **10** 11)55

11 11)121 **12** 11)44 **13** 11)88 **14** 11)11 **15** 11)66

16 11)77 **17** 11)55 **18** 11)132 **19** 11)99 **20** 11)110

Division   ÷11C

1 $11\overline{)132}$ **2** $11\overline{)77}$ **3** $11\overline{)121}$ **4** $11\overline{)33}$ **5** $11\overline{)99}$

6 $11\overline{)11}$ **7** $11\overline{)110}$ **8** $11\overline{)88}$ **9** $11\overline{)44}$ **10** $11\overline{)66}$

11 $11\overline{)77}$ **12** $11\overline{)33}$ **13** $11\overline{)55}$ **14** $11\overline{)22}$ **15** $11\overline{)121}$

16 $11\overline{)44}$ **17** $11\overline{)99}$ **18** $11\overline{)132}$ **19** $11\overline{)88}$ **20** $11\overline{)110}$

Goal Time _____ Actual Time _____ Score _____

Division Challenge ÷11

1 11)22 **2** 11)198 **3** 11)99 **4** 11)154 **5** 11)44

6 11)165 **7** 11)88 **8** 11)11 **9** 11)66 **10** 11)220

11 11)121 **12** 11)33 **13** 11)176 **14** 11)187 **15** 11)110

16 11)77 **17** 11)143 **18** 11)209 **19** 11)55 **20** 11)132

Division ÷12A

1 12)12 **2** 12)120 **3** 12)72 **4** 12)36 **5** 12)96

6 12)60 **7** 12)24 **8** 12)84 **9** 12)108 **10** 12)48

11 12)132 **12** 12)36 **13** 12)144 **14** 12)96 **15** 12)12

16 12)108 **17** 12)48 **18** 12)72 **19** 12)24 **20** 12)60

Goal Time _____ Actual Time _____ Score _____

Division ÷12B

1 12)60   **2** 12)24   **3** 12)96   **4** 12)144   **5** 12)48

6 12)108   **7** 12)132   **8** 12)36   **9** 12)120   **10** 12)72

11 12)84   **12** 12)144   **13** 12)12   **14** 12)48   **15** 12)96

16 12)72   **17** 12)108   **18** 12)24   **19** 12)36   **20** 12)132

Name _____   Date _____

Goal Time _____   Actual Time _____   Score _____

Division   ÷12C

1 12)96   **2** 12)60   **3** 12)144   **4** 12)120   **5** 12)108

6 12)84   **7** 12)12   **8** 12)72   **9** 12)132   **10** 12)48

11 12)24   **12** 12)36   **13** 12)96   **14** 12)60   **15** 12)120

16 12)108   **17** 12)132   **18** 12)12   **19** 12)144   **20** 12)84

Timed Tests: Multiplication and Division © 2004 Creative Teaching Press

Goal Time _____ Actual Time _____ Score _____

Division Challenge   ÷12

1 12)60 **2** 12)108 **3** 12)204 **4** 12)24 **5** 12)168

6 12)144 **7** 12)12 **8** 12)84 **9** 12)228 **10** 12)132

11 12)36 **12** 12)120 **13** 12)180 **14** 12)48 **15** 12)216

16 12)156 **17** 12)72 **18** 12)240 **19** 12)96 **20** 12)192

Name _____ Date _____

Goal Time _____ Actual Time _____ Score _____

 **Practice Division**   ÷10–12A

1 12)‾36   **2** 10)‾100   **3** 11)‾33   **4** 10)‾40   **5** 12)‾48

6 10)‾60   **7** 11)‾99   **8** 12)‾60   **9** 11)‾11   **10** 12)‾132

11 11)‾77   **12** 12)‾84   **13** 10)‾20   **14** 12)‾144   **15** 11)‾55

16 10)‾120   **17** 11)‾44   **18** 12)‾108   **19** 10)‾80   **20** 11)‾121

Timed Tests: Multiplication and Division © 2006 Creative Teaching Press

Name _____ Date _____

Goal Time _____ Actual Time _____ Score _____

 **Practice Division**   ÷10−12B

1 11)55   **2** 10)30   **3** 11)88   **4** 12)120   **5** 10)70

6 10)90   **7** 11)66   **8** 12)72   **9** 11)22   **10** 12)48

11 10)110   **12** 12)24   **13** 11)44   **14** 10)10   **15** 11)110

16 12)144   **17** 10)50   **18** 12)96   **19** 11)132   **20** 10)60

Division Review   ÷1–12A

1 $9\overline{)36}$ **2** $4\overline{)20}$ **3** $2\overline{)16}$ **4** $1\overline{)11}$ **5** $6\overline{)48}$

6 $7\overline{)21}$ **7** $5\overline{)50}$ **8** $8\overline{)48}$ **9** $3\overline{)21}$ **10** $10\overline{)30}$

11 $3\overline{)9}$ **12** $4\overline{)44}$ **13** $2\overline{)24}$ **14** $8\overline{)96}$ **15** $6\overline{)36}$

16 $11\overline{)66}$ **17** $9\overline{)81}$ **18** $7\overline{)42}$ **19** $5\overline{)35}$ **20** $12\overline{)144}$

Name _____ Date _____

Goal Time _____ Actual Time _____ Score _____

 **Division Review** ÷1−12B

1 5)‾25‾ **2** 9)‾63‾ **3** 7)‾35‾ **4** 10)‾50‾ **5** 2)‾14‾

6 3)‾18‾ **7** 1)‾5‾ **8** 12)‾96‾ **9** 4)‾28‾ **10** 11)‾77‾

11 11)‾121‾ **12** 6)‾42‾ **13** 5)‾40‾ **14** 3)‾33‾ **15** 8)‾72‾

16 4)‾36‾ **17** 8)‾56‾ **18** 10)‾120‾ **19** 7)‾49‾ **20** 12)‾48‾

 # Multiplication and Division Review A

1 11)99 **2** $\begin{array}{r} 2 \\ \times\ 7 \\ \hline \end{array}$ **3** 5)40 **4** $\begin{array}{r} 7 \\ \times\ 8 \\ \hline \end{array}$ **5** 3)15

6 12)48 **7** $\begin{array}{r} 3 \\ \times\ 4 \\ \hline \end{array}$ **8** $\begin{array}{r} 10 \\ \times\ 6 \\ \hline \end{array}$ **9** 8)64 **10** $\begin{array}{r} 4 \\ \times\ 9 \\ \hline \end{array}$

11 $\begin{array}{r} 9 \\ \times\ 7 \\ \hline \end{array}$ **12** 7)35 **13** $\begin{array}{r} 11 \\ \times\ 8 \\ \hline \end{array}$ **14** 2)12 **15** $\begin{array}{r} 6 \\ \times\ 6 \\ \hline \end{array}$

16 4)28 **17** $\begin{array}{r} 12 \\ \times\ 9 \\ \hline \end{array}$ **18** $\begin{array}{r} 5 \\ \times 12 \\ \hline \end{array}$ **19** 9)72 **20** $\begin{array}{r} 8 \\ \times\ 4 \\ \hline \end{array}$

Multiplication and Division Review B

1 $6\overline{)54}$ **2** $10\overline{)70}$ **3** $\begin{array}{r} 3 \\ \times\ 8 \\ \hline \end{array}$ **4** $8\overline{)32}$ **5** $4\overline{)24}$

6 $\begin{array}{r} 11 \\ \times\ 5 \\ \hline \end{array}$ **7** $\begin{array}{r} 7 \\ \times\ 3 \\ \hline \end{array}$ **8** $5\overline{)60}$ **9** $\begin{array}{r} 9 \\ \times\ 5 \\ \hline \end{array}$ **10** $\begin{array}{r} 2 \\ \times\ 9 \\ \hline \end{array}$

11 $6\overline{)42}$ **12** $3\overline{)21}$ **13** $\begin{array}{r} 8 \\ \times\ 6 \\ \hline \end{array}$ **14** $\begin{array}{r} 4 \\ \times 11 \\ \hline \end{array}$ **15** $11\overline{)55}$

16 $2\overline{)10}$ **17** $9\overline{)81}$ **18** $7\overline{)49}$ **19** $\begin{array}{r} 10 \\ \times 11 \\ \hline \end{array}$ **20** $\begin{array}{r} 5 \\ \times\ 6 \\ \hline \end{array}$

 # Multiplication and Division Review C

① 12)144   **②** 6 × 9   **③** 8)40   **④** 9 × 3   **⑤** 10)100

⑥ 2 × 4   **⑦** 11)121   **⑧** 4 × 5   **⑨** 4)32   **⑩** 7)84

⑪ 7 × 9   **⑫** 9)108   **⑬** 6)66   **⑭** 12 × 7   **⑮** 7)63

⑯ 10 × 8   **⑰** 11 × 4   **⑱** 3)9   **⑲** 8 × 6   **⑳** 5)45

 # Multiplication and Division Review D

1 $\begin{array}{r} 8 \\ \times\ 2 \\ \hline \end{array}$

2 $5\overline{)60}$

3 $\begin{array}{r} 9 \\ \times\ 6 \\ \hline \end{array}$

4 $3\overline{)36}$

5 $\begin{array}{r} 11 \\ \times\ 3 \\ \hline \end{array}$

6 $\begin{array}{r} 10 \\ \times\ 5 \\ \hline \end{array}$

7 $7\overline{)56}$

8 $\begin{array}{r} 6 \\ \times\ 4 \\ \hline \end{array}$

9 $12\overline{)132}$

10 $\begin{array}{r} 3 \\ \times\ 6 \\ \hline \end{array}$

11 $2\overline{)14}$

12 $\begin{array}{r} 5 \\ \times\ 3 \\ \hline \end{array}$

13 $11\overline{)66}$

14 $8\overline{)72}$

15 $6\overline{)48}$

16 $\begin{array}{r} 12 \\ \times\ 6 \\ \hline \end{array}$

17 $10\overline{)90}$

18 $4\overline{)16}$

19 $\begin{array}{r} 7 \\ \times\ 4 \\ \hline \end{array}$

20 $9\overline{)54}$

Multiplication and Division Review E

1 $9\overline{)36}$ **2** $\begin{array}{r}4\\\times12\end{array}$ **3** $5\overline{)35}$ **4** $\begin{array}{r}11\\\times12\end{array}$ **5** $3\overline{)24}$

6 $\begin{array}{r}6\\\times\,2\end{array}$ **7** $\begin{array}{r}12\\\times\,3\end{array}$ **8** $11\overline{)88}$ **9** $\begin{array}{r}5\\\times\,7\end{array}$ **10** $8\overline{)96}$

11 $4\overline{)48}$ **12** $\begin{array}{r}8\\\times\,9\end{array}$ **13** $12\overline{)84}$ **14** $\begin{array}{r}10\\\times\,9\end{array}$ **15** $9\overline{)45}$

16 $2\overline{)8}$ **17** $\begin{array}{r}12\\\times\,6\end{array}$ **18** $6\overline{)36}$ **19** $\begin{array}{r}9\\\times\,9\end{array}$ **20** $7\overline{)42}$

 **Multiplication and Division Review F**

1 $\begin{array}{r} 3 \\ \times\ 9 \\ \hline \end{array}$ **2** $4\overline{)36}$ **3** $\begin{array}{r} 10 \\ \times\ 7 \\ \hline \end{array}$ **4** $9\overline{)63}$ **5** $6\overline{)72}$

6 $10\overline{)110}$ **7** $\begin{array}{r} 5 \\ \times\ 5 \\ \hline \end{array}$ **8** $7\overline{)35}$ **9** $\begin{array}{r} 6 \\ \times\ 7 \\ \hline \end{array}$ **10** $\begin{array}{r} 12 \\ \times\ 8 \\ \hline \end{array}$

11 $\begin{array}{r} 7 \\ \times\ 7 \\ \hline \end{array}$ **12** $8\overline{)24}$ **13** $\begin{array}{r} 11 \\ \times 11 \\ \hline \end{array}$ **14** $5\overline{)30}$ **15** $\begin{array}{r} 12 \\ \times\ 2 \\ \hline \end{array}$

16 $12\overline{)108}$ **17** $\begin{array}{r} 12 \\ \times\ 9 \\ \hline \end{array}$ **18** $11\overline{)77}$ **19** $\begin{array}{r} 4 \\ \times\ 8 \\ \hline \end{array}$ **20** $3\overline{)18}$

Answer Key

×2A
1. 16	2. 6	3. 0	4. 8	5. 12
6. 24	7. 2	8. 14	9. 10	10. 4
11. 10	12. 18	13. 8	14. 4	15. 22
16. 2	17. 6	18. 20	19. 12	20. 14

×2B
1. 2	2. 10	3. 4	4. 18	5. 6
6. 8	7. 12	8. 20	9. 24	10. 16
11. 24	12. 0	13. 14	14. 12	15. 4
16. 16	17. 6	18. 18	19. 22	20. 20

×2C
1. 4	2. 14	3. 20	4. 6	5. 16
6. 18	7. 0	8. 10	9. 2	10. 8
11. 20	12. 8	13. 16	14. 14	15. 22
16. 10	17. 6	18. 24	19. 18	20. 12

Challenge ×2
1. 24	2. 0	3. 14	4. 8	5. 20
6. 6	7. 18	8. 1	9. 11	10. 10
11. 9	12. 2	13. 16	14. 4	15. 12
16. 11	17. 3	18. 0	19. 7	20. 24

×3A
1. 6	2. 9	3. 18	4. 21	5. 27
6. 15	7. 3	8. 36	9. 9	10. 0
11. 12	12. 18	13. 21	14. 30	15. 6
16. 24	17. 33	18. 12	19. 15	20. 3

×3B
1. 33	2. 3	3. 15	4. 24	5. 9
6. 24	7. 18	8. 30	9. 36	10. 21
11. 15	12. 0	13. 6	14. 12	15. 27
16. 36	17. 27	18. 21	19. 9	20. 18

×3C
1. 9	2. 15	3. 21	4. 6	5. 30
6. 24	7. 12	8. 33	9. 0	10. 36
11. 21	12. 3	13. 15	14. 27	15. 9
16. 18	17. 24	18. 30	19. 36	20. 33

Challenge ×3
1. 4	2. 8	3. 6	4. 0	5. 27
6. 2	7. 15	8. 10	9. 33	10. 5
11. 6	12. 9	13. 24	14. 1	15. 12
16. 11	17. 18	18. 3	19. 7	20. 21

Practice ×0–3A
1. 24	2. 0	3. 9	4. 0	5. 6
6. 0	7. 16	8. 9	9. 22	10. 3
11. 0	12. 14	13. 0	14. 27	15. 7
16. 12	17. 12	18. 4	19. 12	20. 15

Practice ×0–3B
1. 0	2. 14	3. 0	4. 18	5. 6
6. 5	7. 33	8. 6	9. 4	10. 11
11. 8	12. 0	13. 16	14. 0	15. 20
16. 0	17. 21	18. 30	19. 18	20. 10

×4A
1. 20	2. 36	3. 4	4. 8	5. 24
6. 48	7. 28	8. 12	9. 40	10. 0
11. 32	12. 8	13. 0	14. 28	15. 36
16. 40	17. 16	18. 48	19. 12	20. 44

×4B
1. 44	2. 4	3. 16	4. 20	5. 24
6. 0	7. 24	8. 48	9. 12	10. 28
11. 20	12. 4	13. 28	14. 32	15. 36
16. 16	17. 8	18. 40	19. 44	20. 24

×4C
1. 0	2. 40	3. 28	4. 12	5. 20
6. 24	7. 8	8. 32	9. 8	10. 44
11. 36	12. 48	13. 40	14. 16	15. 20
16. 24	17. 32	18. 4	19. 0	20. 44

Challenge ×4
1. 4	2. 6	3. 12	4. 1	5. 36
6. 11	7. 0	8. 8	9. 40	10. 2
11. 10	12. 12	13. 8	14. 5	15. 0
16. 9	17. 20	18. 3	19. 7	20. 48

×5A
1. 10	2. 40	3. 30	4. 20	5. 45
6. 35	7. 50	8. 0	9. 55	10. 15
11. 25	12. 45	13. 60	14. 15	15. 40
16. 0	17. 5	18. 35	19. 30	20. 50

×5B
1. 25	2. 45	3. 10	4. 15	5. 60
6. 5	7. 55	8. 35	9. 20	10. 50
11. 20	12. 30	13. 45	14. 25	15. 40
16. 55	17. 10	18. 60	19. 30	20. 5

×5C
1. 50	2. 20	3. 60	4. 40	5. 15
6. 35	7. 25	8. 10	9. 45	10. 30
11. 55	12. 5	13. 40	14. 60	15. 30
16. 15	17. 35	18. 10	19. 0	20. 45

Challenge ×5
1. 9	2. 40	3. 11	4. 50	5. 5
6. 10	7. 3	8. 55	9. 12	10. 0
11. 2	12. 30	13. 4	14. 45	15. 7
16. 6	17. 35	18. 8	19. 1	20. 10

×6A
1. 6	2. 60	3. 24	4. 48	5. 18
6. 48	7. 30	8. 54	9. 12	10. 66
11. 54	12. 6	13. 66	14. 60	15. 36
16. 18	17. 36	18. 42	19. 24	20. 72

×6B
1. 36	2. 60	3. 12	4. 72	5. 42
6. 6	7. 54	8. 24	9. 30	10. 66
11. 48	12. 18	13. 42	14. 36	15. 0
16. 30	17. 72	18. 6	19. 54	20. 24

×6C
1. 18	2. 66	3. 30	4. 72	5. 54
6. 48	7. 6	8. 12	9. 42	10. 0
11. 36	12. 60	13. 24	14. 18	15. 12
16. 24	17. 72	18. 48	19. 30	20. 60

Challenge ×6
1. 9	2. 30	3. 2	4. 66	5. 6
6. 12	7. 4	8. 42	9. 11	10. 1
11. 8	12. 30	13. 5	14. 24	15. 3
16. 10	17. 54	18. 0	19. 7	20. 18

Practice ×4–6A
1. 24	2. 40	3. 45	4. 36	5. 20
6. 12	7. 35	8. 66	9. 36	10. 48
11. 10	12. 54	13. 48	14. 18	15. 32
16. 42	17. 8	18. 0	19. 28	20. 25

Practice ×4–6B
1. 15	2. 24	3. 30	4. 45	5. 50
6. 16	7. 60	8. 44	9. 42	10. 4
11. 32	12. 72	13. 12	14. 36	15. 55
16. 20	17. 6	18. 25	19. 60	20. 54

×7A
1. 28	2. 56	3. 84	4. 14	5. 49
6. 42	7. 70	8. 21	9. 28	10. 0
11. 21	12. 14	13. 56	14. 35	15. 63
16. 49	17. 35	18. 42	19. 77	20. 70

×7B
1. 35	2. 70	3. 21	4. 14	5. 84
6. 49	7. 0	8. 28	9. 63	10. 56
11. 21	12. 77	13. 35	14. 42	15. 7
16. 63	17. 56	18. 0	19. 7	20. 77

×7C
1. 49	2. 28	3. 21	4. 63	5. 0
6. 7	7. 35	8. 42	9. 77	10. 56
11. 70	12. 14	13. 35	14. 49	15. 63
16. 42	17. 28	18. 56	19. 21	20. 84

Challenge ×7

1. 3 2. 49 3. 10 4. 14 5. 6
6. 5 7. 11 8. 56 9. 1 10. 9
11. 8 12. 63 13. 12 14. 21 15. 4
16. 7 17. 84 18. 0 19. 2 20. 70

×8A

1. 0 2. 32 3. 96 4. 72 5. 16
6. 64 7. 8 8. 24 9. 48 10. 40
11. 24 12. 80 13. 40 14. 0 15. 72
16. 8 17. 56 18. 64 19. 16 20. 88

×8B

1. 80 2. 24 3. 48 4. 8 5. 88
6. 96 7. 40 8. 32 9. 16 10. 56
11. 16 12. 56 13. 88 14. 24 15. 96
16. 0 17. 72 18. 80 19. 32 20. 48

×8C

1. 24 2. 32 3. 0 4. 72 5. 48
6. 40 7. 64 8. 16 9. 88 10. 56
11. 8 12. 96 13. 80 14. 24 15. 40
16. 72 17. 48 18. 32 19. 64 20. 80

Challenge ×8

1. 7 2. 2 3. 56 4. 11 5. 3
6. 8 7. 72 8. 4 9. 40 10. 9
11. 1 12. 96 13. 6 14. 24 15. 10
16. 12 17. 32 18. 0 19. 5 20. 64

×9A

1. 36 2. 72 3. 0 4. 99 5. 27
6. 108 7. 45 8. 18 9. 90 10. 9
11. 63 12. 36 13. 27 14. 81 15. 0
16. 72 17. 90 18. 81 19. 54 20. 45

×9B

1. 99 2. 9 3. 36 4. 27 5. 54
6. 72 7. 45 8. 18 9. 0 10. 63
11. 18 12. 90 13. 108 14. 81 15. 45
16. 63 17. 108 18. 9 19. 36 20. 99

×9C

1. 27 2. 81 3. 45 4. 90 5. 72
6. 63 7. 54 8. 18 9. 0 10. 108
11. 36 12. 99 13. 54 14. 72 15. 63
16. 45 17. 27 18. 108 19. 9 20. 81

Challenge ×9

1. 5 2. 81 3. 11 4. 18 5. 4
6. 8 7. 90 8. 1 9. 63 10. 6
11. 2 12. 72 13. 10 14. 0 15. 3
16. 7 17. 54 18. 12 19. 9 20. 45

Practice ×7–9A

1. 35 2. 18 3. 28 4. 40 5. 84
6. 54 7. 32 8. 70 9. 48 10. 35
11. 80 12. 88 13. 72 14. 56 15. 108
16. 14 17. 36 18. 45 19. 24 20. 90

Practice ×7–9B

1. 21 2. 64 3. 27 4. 77 5. 45
6. 32 7. 99 8. 42 9. 40 10. 49
11. 81 12. 16 13. 63 14. 96 15. 9
16. 48 17. 35 18. 72 19. 56 20. 36

×10A

1. 80 2. 40 3. 30 4. 120 5. 70
6. 100 7. 10 8. 0 9. 20 10. 50
11. 60 12. 80 13. 110 14. 20 15. 100
16. 50 17. 40 18. 70 19. 90 20. 30

×10B

1. 20 2. 110 3. 40 4. 90 5. 100
6. 120 7. 60 8. 80 9. 10 10. 0
11. 70 12. 10 13. 20 14. 30 15. 50
16. 40 17. 90 18. 120 19. 110 20. 60

×10C

1. 70 2. 50 3. 40 4. 110 5. 20
6. 30 7. 120 8. 60 9. 90 10. 80
11. 20 12. 90 13. 120 14. 50 15. 110
16. 100 17. 70 18. 10 19. 0 20. 40

Challenge ×10

1. 80 2. 5 3. 30 4. 9 5. 20
6. 7 7. 60 8. 4 9. 12 10. 2
11. 0 12. 1 13. 50 14. 8 15. 11
16. 10 17. 3 18. 90 19. 6 20. 40

×11A

1. 33 2. 99 3. 77 4. 132 5. 0
6. 66 7. 55 8. 11 9. 110 10. 88
11. 22 12. 88 13. 121 14. 33 15. 44
16. 132 17. 77 18. 66 19. 99 20. 55

×11B

1. 66 2. 44 3. 11 4. 99 5. 55
6. 33 7. 121 8. 88 9. 110 10. 132
11. 110 12. 55 13. 0 14. 77 15. 22
16. 88 17. 99 18. 132 19. 44 20. 121

×11C

1. 22 2. 66 3. 132 4. 88 5. 99
6. 77 7. 110 8. 11 9. 33 10. 44
11. 55 12. 121 13. 110 14. 22 15. 77
16. 33 17. 99 18. 66 19. 44 20. 132

Challenge ×11

1. 6 2. 8 3. 99 4. 3 5. 22
6. 88 7. 5 8. 44 9. 11 10. 77
11. 12 12. 66 13. 2 14. 0 15. 9
16. 7 17. 1 18. 121 19. 4 20. 10

×12A

1. 36 2. 60 3. 0 4. 108 5. 48
6. 132 7. 24 8. 96 9. 84 10. 72
11. 108 12. 36 13. 60 14. 12 15. 120
16. 84 17. 48 18. 72 19. 96 20. 144

×12B

1. 24 2. 72 3. 48 4. 144 5. 84
6. 132 7. 120 8. 96 9. 84 10. 72
11. 12 12. 60 13. 0 14. 108 15. 120
16. 48 17. 24 18. 36 19. 144 20. 132

×12C

1. 60 2. 132 3. 36 4. 48 5. 96
6. 12 7. 108 8. 120 9. 72 10. 0
11. 144 12. 84 13. 132 14. 24 15. 120
16. 72 17. 12 18. 48 19. 96 20. 60

Challenge ×12

1. 96 2. 2 3. 48 4. 5 5. 36
6. 9 7. 108 8. 6 9. 12 10. 8
11. 11 12. 84 13. 72 14. 3 15. 10
16. 4 17. 1 18. 60 19. 7 20. 132

Practice ×10–12A

1. 60 2. 88 3. 60 4. 108 5. 70
6. 36 7. 40 8. 77 9. 50 10. 120
11. 66 12. 80 13. 99 14. 24 15. 55
16. 36 17. 90 18. 144 19. 33 20. 72

Practice ×10–12B

1. 24 2. 10 3. 44 4. 84 5. 90
6. 121 7. 36 8. 12 9. 20 10. 88
11. 30 12. 22 13. 48 14. 110 15. 120
16. 99 17. 80 18. 70 19. 96 20. 0

Review ×1–12A

1. 36 2. 12 3. 56 4. 81 5. 16
6. 27 7. 66 8. 49 9. 100 10. 60
11. 55 12. 16 13. 132 14. 14 15. 32
16. 30 17. 15 18. 72 19. 24 20. 96

Review ×1–12B

1. 84 2. 18 3. 48 4. 40 5. 9
6. 22 7. 25 8. 36 9. 21 10. 88
11. 77 12. 60 13. 28 14. 110 15. 144
16. 64 17. 20 18. 18 19. 40 20. 54

÷2A

1. 6	2. 2	3. 6	4. 1	5. 3
6. 10	7. 12	8. 8	9. 4	10. 7
11. 5	12. 3	13. 9	14. 6	15. 12
16. 4	17. 11	18. 2	19. 8	20. 5

Practice ÷1–3B

1. 3	2. 6	3. 2	4. 11	5. 3
6. 11	7. 5	8. 2	9. 12	10. 5
11. 6	12. 8	13. 9	14. 1	15. 9
16. 2	17. 6	18. 10	19. 1	20. 11

÷6A

1. 4	2. 10	3. 6	4. 2	5. 8
6. 1	7. 9	8. 3	9. 11	10. 5
11. 7	12. 2	13. 12	14. 10	15. 6
16. 3	17. 8	18. 9	19. 4	20. 1

÷2B

1. 11	2. 2	3. 8	4. 5	5. 6
6. 10	7. 9	8. 4	9. 12	10. 7
11. 5	12. 6	13. 11	14. 3	15. 10
16. 1	17. 7	18. 2	19. 8	20. 9

÷4A

1. 4	2. 10	3. 2	4. 6	5. 8
6. 1	7. 7	8. 11	9. 5	10. 9
11. 8	12. 2	13. 3	14. 12	15. 7
16. 6	17. 5	18. 9	19. 4	20. 1

÷6B

1. 1	2. 9	3. 12	4. 3	5. 10
6. 5	7. 11	8. 8	9. 7	10. 4
11. 6	12. 2	13. 10	14. 9	15. 12
16. 11	17. 7	18. 4	19. 8	20. 5

÷2C

1. 11	2. 2	3. 8	4. 5	5. 6
6. 10	7. 9	8. 4	9. 12	10. 7
11. 1	12. 3	13. 5	14. 11	15. 2
16. 9	17. 12	18. 10	19. 7	20. 4

÷4B

1. 1	2. 3	3. 7	4. 5	5. 12
6. 10	7. 6	8. 4	9. 8	10. 2
11. 7	12. 9	13. 11	14. 1	15. 3
16. 8	17. 4	18. 7	19. 10	20. 5

÷6C

1. 5	2. 3	3. 9	4. 7	5. 12
6. 6	7. 10	8. 4	9. 8	10. 2
11. 1	12. 12	13. 11	14. 5	15. 3
16. 2	17. 8	18. 9	19. 6	20. 4

Challenge ÷2

1. 12	2. 3	3. 8	4. 15	5. 19
6. 10	7. 17	8. 14	9. 2	10. 7
11. 9	12. 6	13. 20	14. 11	15. 1
16. 4	17. 18	18. 5	19. 16	20. 13

÷4C

1. 5	2. 12	3. 6	4. 3	5. 9
6. 2	7. 8	8. 4	9. 10	10. 1
11. 10	12. 7	13. 3	14. 11	15. 6
16. 12	17. 9	18. 4	19. 5	20. 2

Challenge ÷6

1. 4	2. 18	3. 9	4. 6	5. 12
6. 11	7. 2	8. 15	9. 19	10. 5
11. 7	12. 17	13. 1	14. 13	15. 10
16. 14	17. 8	18. 20	19. 3	20. 16

÷3A

1. 2	2. 10	3. 3	4. 4	5. 5
6. 7	7. 9	8. 1	9. 6	10. 8
11. 5	12. 11	13. 10	14. 3	15. 1
16. 12	17. 4	18. 6	19. 2	20. 9

Challenge ÷4

1. 16	2. 6	3. 18	4. 10	5. 14
6. 5	7. 19	8. 1	9. 9	10. 3
11. 15	12. 11	13. 20	14. 7	15. 17
16. 8	17. 13	18. 2	19. 4	20. 12

Practice ÷4–6A

1. 4	2. 1	3. 11	4. 3	5. 11
6. 1	7. 7	8. 3	9. 2	10. 4
11. 5	12. 11	13. 3	14. 9	15. 1
16. 9	17. 7	18. 5	19. 6	20. 7

÷3B

1. 4	2. 8	3. 10	4. 6	5. 1
6. 3	7. 7	8. 11	9. 2	10. 9
11. 6	12. 5	13. 7	14. 8	15. 4
16. 1	17. 9	18. 2	19. 11	20. 3

÷5A

1. 10	2. 2	3. 7	4. 4	5. 9
6. 5	7. 1	8. 11	9. 6	10. 12
11. 8	12. 9	13. 3	14. 2	15. 7
16. 4	17. 6	18. 10	19. 5	20. 1

Practice ÷4–6B

1. 5	2. 5	3. 5	4. 12	5. 10
6. 12	7. 2	8. 9	9. 2	10. 6
11. 10	12. 4	13. 4	14. 1	15. 8
16. 8	17. 5	18. 6	19. 10	20. 12

÷3C

1. 1	2. 5	3. 11	4. 7	5. 9
6. 8	7. 3	8. 10	9. 6	10. 2
11. 4	12. 2	13. 9	14. 10	15. 5
16. 8	17. 6	18. 7	19. 1	20. 3

÷5B

1. 1	2. 8	3. 6	4. 3	5. 9
6. 5	7. 12	8. 2	9. 10	10. 7
11. 4	12. 11	13. 5	14. 9	15. 1
16. 2	17. 8	18. 3	19. 6	20. 12

÷7A

1. 4	2. 8	3. 2	4. 11	5. 6
6. 5	7. 1	8. 7	9. 3	10. 9
11. 10	12. 3	13. 12	14. 8	15. 4
16. 11	17. 6	18. 5	19. 1	20. 7

Challenge ÷3

1. 8	2. 18	3. 16	4. 11	5. 6
6. 15	7. 10	8. 3	9. 7	10. 14
11. 5	12. 20	13. 17	14. 12	15. 1
16. 13	17. 19	18. 4	19. 9	20. 2

÷5C

1. 8	2. 11	3. 6	4. 3	5. 10
6. 5	7. 2	8. 9	9. 12	10. 4
11. 10	12. 7	13. 1	14. 2	15. 6
16. 3	17. 9	18. 12	19. 8	20. 5

÷7B

1. 3	2. 5	3. 1	4. 9	5. 7
6. 2	7. 10	8. 6	9. 12	10. 4
11. 8	12. 7	13. 11	14. 5	15. 2
16. 12	17. 4	18. 3	19. 10	20. 6

Practice ÷1–3A

1. 4	2. 7	3. 4	4. 8	5. 8
6. 5	7. 6	8. 10	9. 4	10. 7
11. 12	12. 9	13. 5	14. 9	15. 7
16. 12	17. 10	18. 3	19. 11	20. 7

Challenge ÷5

1. 9	2. 16	3. 5	4. 2	5. 11
6. 14	7. 1	8. 12	9. 7	10. 18
11. 4	12. 8	13. 19	14. 6	15. 17
16. 15	17. 10	18. 20	19. 3	20. 13

÷7C

1. 12	2. 8	3. 4	4. 6	5. 10
6. 3	7. 7	8. 9	9. 2	10. 11
11. 10	12. 6	13. 1	14. 5	15. 12
16. 7	17. 11	18. 3	19. 9	20. 8

Challenge ÷7

1. 1	2. 7	3. 17	4. 12	5. 4
6. 15	7. 9	8. 2	9. 20	10. 13
11. 3	12. 10	13. 18	14. 5	15. 16
16. 11	17. 6	18. 14	19. 19	20. 8

÷8A

1. 5	2. 8	3. 3	4. 7	5. 1
6. 2	7. 10	8. 6	9. 9	10. 4
11. 7	12. 3	13. 11	14. 12	15. 8
16. 4	17. 5	18. 9	19. 2	20. 6

÷8B

1. 4	2. 9	3. 2	4. 11	5. 6
6. 3	7. 7	8. 12	9. 5	10. 8
11. 11	12. 1	13. 6	14. 10	15. 9
16. 8	17. 12	18. 5	19. 7	20. 3

÷8C

1. 8	2. 3	3. 10	4. 5	5. 12
6. 6	7. 9	8. 4	9. 11	10. 1
11. 7	12. 12	13. 2	14. 10	15. 8
16. 11	17. 5	18. 6	19. 9	20. 4

Challenge ÷8

1. 5	2. 12	3. 1	4. 15	5. 8
6. 14	7. 6	8. 18	9. 3	10. 10
11. 2	12. 11	13. 7	14. 13	15. 19
16. 17	17. 4	18. 20	19. 9	20. 16

÷9A

1. 7	2. 1	3. 5	4. 9	5. 3
6. 4	7. 8	8. 10	9. 2	10. 6
11. 3	12. 5	13. 12	14. 7	15. 11
16. 6	17. 2	18. 1	19. 8	20. 10

÷9B

1. 9	2. 2	3. 7	4. 11	5. 6
6. 3	7. 12	8. 5	9. 1	10. 8
11. 11	12. 6	13. 4	14. 9	15. 10
16. 3	17. 4	18. 8	19. 7	20. 12

÷9C

1. 3	2. 6	3. 8	4. 4	5. 11
6. 5	7. 9	8. 7	9. 2	10. 12
11. 4	12. 1	13. 3	14. 10	15. 6
16. 12	17. 9	18. 11	19. 5	20. 7

Challenge ÷9

1. 3	2. 12	3. 17	4. 1	5. 8
6. 2	7. 14	8. 5	9. 18	10. 11
11. 10	12. 4	13. 19	14. 7	15. 13
16. 6	17. 16	18. 9	19. 20	20. 15

Practice ÷7–9A

1. 6	2. 7	3. 5	4. 10	5. 9
6. 6	7. 12	8. 5	9. 2	10. 9
11. 3	12. 11	13. 8	14. 11	15. 3
16. 7	17. 4	18. 4	19. 1	20. 8

Practice ÷7–9B

1. 2	2. 10	3. 8	4. 7	5. 2
6. 9	7. 10	8. 12	9. 1	10. 3
11. 4	12. 6	13. 4	14. 11	15. 4
16. 5	17. 12	18. 8	19. 6	20. 8

÷10A

1. 4	2. 7	3. 2	4. 10	5. 6
6. 12	7. 1	8. 8	9. 3	10. 9
11. 11	12. 6	13. 5	14. 7	15. 2
16. 3	17. 8	18. 9	19. 4	20. 12

÷10B

1. 3	2. 6	3. 11	4. 9	5. 12
6. 8	7. 10	8. 5	9. 2	10. 7
11. 4	12. 7	13. 1	14. 3	15. 11
16. 9	17. 12	18. 8	19. 5	20. 10

÷10C

1. 5	2. 10	3. 3	4. 7	5. 8
6. 9	7. 11	8. 6	9. 4	10. 1
11. 2	12. 8	13. 12	14. 3	15. 5
16. 11	17. 4	18. 7	19. 6	20. 9

Challenge ÷10

1. 7	2. 13	3. 10	4. 2	5. 17
6. 12	7. 1	8. 8	9. 15	10. 4
11. 3	12. 14	13. 6	14. 19	15. 16
16. 11	17. 18	18. 5	19. 20	20. 9

÷11A

1. 5	2. 9	3. 1	4. 11	5. 7
6. 2	7. 8	8. 6	9. 10	10. 3
11. 12	12. 7	13. 4	14. 11	15. 9
16. 3	17. 6	18. 2	19. 8	20. 5

÷11B

1. 6	2. 10	3. 2	4. 8	5. 4
6. 3	7. 9	8. 7	9. 12	10. 5
11. 11	12. 4	13. 8	14. 1	15. 6
16. 7	17. 5	18. 12	19. 9	20. 10

÷11C

1. 12	2. 7	3. 11	4. 3	5. 9
6. 1	7. 10	8. 8	9. 4	10. 6
11. 7	12. 3	13. 5	14. 2	15. 11
16. 4	17. 9	18. 12	19. 8	20. 10

Challenge ÷11

1. 2	2. 18	3. 9	4. 14	5. 4
6. 15	7. 8	8. 1	9. 6	10. 20
11. 11	12. 3	13. 16	14. 17	15. 10
16. 7	17. 13	18. 19	19. 5	20. 12

÷12A

1. 1	2. 10	3. 6	4. 3	5. 8
6. 5	7. 2	8. 7	9. 9	10. 4
11. 11	12. 3	13. 12	14. 8	15. 1
16. 9	17. 4	18. 6	19. 2	20. 5

÷12B

1. 5	2. 2	3. 8	4. 12	5. 4
6. 9	7. 11	8. 3	9. 10	10. 6
11. 7	12. 12	13. 1	14. 4	15. 8
16. 6	17. 9	18. 2	19. 3	20. 11

÷12C

1. 8	2. 5	3. 12	4. 10	5. 9
6. 7	7. 1	8. 6	9. 11	10. 4
11. 2	12. 3	13. 8	14. 5	15. 10
16. 9	17. 11	18. 1	19. 12	20. 7

Challenge ÷12

1. 5	2. 9	3. 17	4. 2	5. 14
6. 12	7. 1	8. 7	9. 19	10. 11
11. 3	12. 10	13. 15	14. 4	15. 18
16. 13	17. 6	18. 20	19. 8	20. 16

Practice ÷10–12A

1. 3	2. 10	3. 3	4. 4	5. 4
6. 6	7. 9	8. 5	9. 1	10. 11
11. 7	12. 7	13. 2	14. 12	15. 5
16. 12	17. 4	18. 9	19. 8	20. 11

Practice ÷10–12B

1. 5	2. 3	3. 8	4. 10	5. 7
6. 9	7. 6	8. 6	9. 2	10. 4
11. 11	12. 2	13. 4	14. 1	15. 10
16. 12	17. 5	18. 8	19. 12	20. 6

Review ÷1–12A

1. 4	2. 5	3. 8	4. 11	5. 8
6. 3	7. 10	8. 6	9. 7	10. 3
11. 3	12. 11	13. 12	14. 12	15. 6
16. 6	17. 9	18. 6	19. 7	20. 12

Review ÷1–12B

1. 5	2. 7	3. 5	4. 5	5. 7
6. 6	7. 5	8. 8	9. 7	10. 7
11. 11	12. 7	13. 8	14. 11	15. 9
16. 9	17. 7	18. 12	19. 7	20. 4

Review × and ÷ A

1. 9	2. 14	3. 8	4. 56	5. 5
6. 4	7. 12	8. 60	9. 8	10. 36
11. 63	12. 5	13. 88	14. 6	15. 36
16. 7	17. 108	18. 60	19. 8	20. 32

Review × and ÷ B

1. 9	2. 7	3. 24	4. 4	5. 6
6. 55	7. 21	8. 12	9. 45	10. 18
11. 7	12. 7	13. 48	14. 44	15. 5
16. 5	17. 9	18. 7	19. 110	20. 30

Review × and ÷ C

1. 12	2. 54	3. 5	4. 27	5. 10
6. 8	7. 11	8. 20	9. 8	10. 12
11. 63	12. 12	13. 11	14. 84	15. 9
16. 80	17. 44	18. 3	19. 48	20. 9

Review × and ÷ D

1. 16	2. 12	3. 54	4. 12	5. 33
6. 50	7. 8	8. 24	9. 11	10. 18
11. 7	12. 15	13. 6	14. 9	15. 8
16. 72	17. 9	18. 4	19. 28	20. 6

Review × and ÷ E

1. 4	2. 48	3. 7	4. 132	5. 8
6. 12	7. 36	8. 8	9. 35	10. 12
11. 12	12. 72	13. 7	14. 90	15. 5
16. 4	17. 72	18. 6	19. 81	20. 6

Review × and ÷ F

1. 27	2. 9	3. 70	4. 7	5. 12
6. 11	7. 25	8. 5	9. 42	10. 96
11. 49	12. 3	13. 121	14. 6	15. 24
16. 9	17. 108	18. 7	19. 32	20. 6